Stop!!!

This program cannot be used effectively without online registration.

Without online registration, you won't have access to practice tests, answer sheets, detailed explanations, automated scoring, personalized study plans.

Before moving forward, please complete your registration using the information given below.

Complete Your Registration Now

Scan the QR Code or Visit

lumoslearning.com/a/tedbooks

Use Access Code: MSTEPG6M-27371-P

Lumos Learning

Step Up Your Skills

How to Use the Lumos Program?

Congratulations on choosing the Lumos M-STEP Program! This isn't just a book. It's a comprehensive program designed to help your child succeed on the state test.

Here is how to use the program:

1. Login to your child's account at ***lumoslearning.com*** using the details sent to your registered email.
2. Have your child take the assigned online Practice Test 1.
3. After finishing the test, click on the *View Study Plan* button on the home page.
4. For each lesson in the study plan, have your child look at the questions in the printed book and answer them in the online answer sheet.
5. After completing all the lessons in the study plan, have your child take Practice Test 2.
6. View your child's performance by going to the *My Reports* section in the top right corner of the home page.

Table of Contents

Chapter 1

Ratios & Proportional Relationships

Do NOT write your answers in this book. To open the answer sheet, scan the QR code or visit ***lumoslearning.com/a/6m001***

Lesson 1: Expressing Ratios

1. A school has an enrollment of 600 students. 330 of the students are girls. Express the fraction of students who are boys in simplest terms.

Ⓐ $\frac{12}{20}$

Ⓑ $\frac{11}{20}$

Ⓒ $\frac{9}{20}$

Ⓓ $\frac{13}{20}$

Boys equal = 270

$\frac{270}{600} = \frac{27}{60} = \frac{9}{20}$

2. In the 14th century, the Sultan of Brunei noticed that his ratio of emeralds to rubies was the same as the ratio of diamonds to pearls. If he had 85 emeralds, 119 rubies, and 45 diamonds, how many pearls did he have?

Ⓐ 17
Ⓑ 22
Ⓒ 58
Ⓓ 63

3. Mr. Fullingham has 75 geese and 125 turkeys. What is the ratio of the number of geese to the total number of birds in simplest terms?

Ⓐ 75:200
Ⓑ 3:8
Ⓒ 125:200
Ⓓ 5:8

4. **The little league team called the Hawks has 7 brunettes, 5 blonds, and 2 redheads. What is the ratio of redheads to the entire team in simplest terms?**

Ⓐ 2:7
Ⓑ 2:5
Ⓒ 2:12
Ⓓ 1:7

5. **The little league team called the Hawks has 7 brunettes, 5 blonds, and 2 redheads. The entire little league division that the Hawks belong to has the same ratio of redheads to everyone else. What is the total number of redheads in that division if the total number of players is 126?**

Ⓐ 9
Ⓑ 14
Ⓒ 18
Ⓓ 24

6. **Barnaby decided to count the number of ducks and geese flying south for the winter. The first day he counted 175 ducks and 63 geese. What is the ratio of ducks to the total number of birds flying overhead in simplest terms?**

Ⓐ 175:63
Ⓑ 175:238
Ⓒ 25:9
Ⓓ 25:34

7. **Barnaby decided to count the number of ducks and geese flying south for the winter. The first day he counted 175 ducks and 63 geese. By the end of migration, Barnaby had counted 4,725 geese. If the ratio of ducks to geese remained the same (175 to 63), how many ducks did he count?**

Ⓐ 13,125
Ⓑ 17,850
Ⓒ 10,695
Ⓓ 14,750

8. **Barbara was baking a cake and could not find her tablespoon measure. The recipe calls for $3\frac{1}{3}$ tablespoons. Each table spoon measure 3 teaspoon. How many teaspoons must Barbara use in order to have the recipe turn out all right?**

Ⓐ 3
Ⓑ 6
Ⓒ 9
Ⓓ 10

9. The ratio of girls to boys in a grade is 6 to 5. If there are 24 girls in the grade then how many students are there altogether?

Ⓐ 14
Ⓑ 24
Ⓒ 34
Ⓓ 44

10. The ratio of pencils to pens in a box is 3 to 2. If there are 30 pencils and pens altogether, how many pencils are there?

Ⓐ 16
Ⓑ 17
Ⓒ 18
Ⓓ 19

11. Which of the following correctly expresses the ratio of shaded bows to the number of total bows? Select all answers that apply.

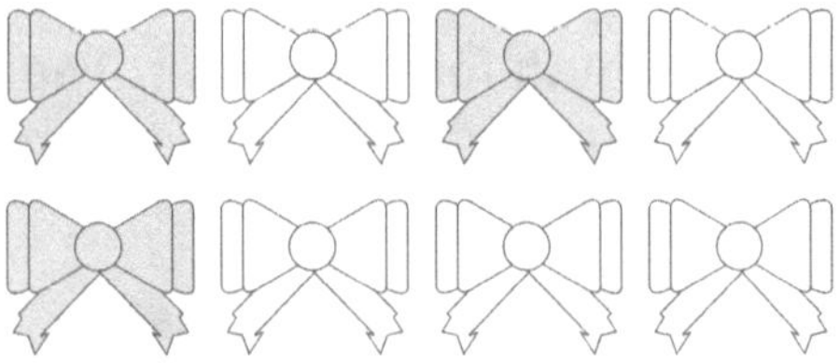

Ⓐ 3:8
Ⓑ 5:8
Ⓒ $\frac{3}{5}$
Ⓓ $\frac{3}{8}$
Ⓔ $\frac{5}{8}$

12. Write the ratio that correctly describes the number of white stars compared to the number of gray stars. Write your answer in the box below.

5 : 9

13. Complete the following table by filling in the blanks with a number that shows the correct ratio that is equivalent to the one shown in the first row.

1	2
2	4
3	6
4	8
5	10
6	12

Do NOT write your answers in this book. To open the answer sheet, scan the QR code or visit ***lumoslearning.com/a/6m002***

Chapter 1 → Lesson 2: Unit Rates

1. Which is a better price: 5 for $1.00, 4 for 85¢, 2 for 25¢, or 6 for $1.10?

Ⓐ 5 for $1.00
Ⓑ 4 for 85¢
Ⓒ 2 for 25¢
Ⓓ 6 for $1.10

2. At grocery Store A, 5 cans of baked beans cost $3.45. At grocery Store B, 7 cans of baked beans cost $5.15. At grocery Store C, 4 cans of baked beans cost $2.46. At grocery Store D, 6 cans of baked beans cost $4.00. How much money would you save if you bought 20 cans of baked beans from grocery store C than if you bought 20 cans of baked beans from grocery store A?

Ⓐ $1.75
Ⓑ $1.25
Ⓒ $1.50
Ⓓ 95¢

3. Beverly drove from Atlantic City to Newark. She drove for 284 miles at a constant speed of 58 mph. How long did it take Beverly to complete the trip?

Ⓐ 4 hours and 45 minutes
Ⓑ 4 hours and 54 minutes
Ⓒ 4 hours and 8 minutes
Ⓓ 4 hours and 89 minutes

4. Don has two jobs. For Job 1, he earns $7.55 an hour. For Job 2, he earns $8.45 an hour. Last week he worked at the first job for 10 hours and at the second job for 15 hours. What were his average earnings per hour?

Ⓐ $8.00
Ⓑ $8.09
Ⓒ $8.15
Ⓓ $8.13

5. It took Marjorie 15 minutes to drive from her house to her daughter's school. If the school was 4 miles away from her house, what was her unit rate of speed?

Ⓐ 16 mph
Ⓑ 8 mph
Ⓒ 4 mph
Ⓓ 30 mph

6. The Belmont race track known as "Big Sandy" is 1½ miles long. In 1973, Secretariat won the Belmont Stakes race in 2 minutes and 30 seconds. Assuming he ran on "Big Sandy", what was his unit speed?

Ⓐ 30 mph
Ⓑ 40 mph
Ⓒ 36 mph
Ⓓ 38 mph

7. If 1 pound of chocolate creams at Philadelphia Candies costs $7.52. How much does that candy cost by the ounce?

Ⓐ 48¢ per oz.
Ⓑ 47¢ per oz.
Ⓒ 75.2¢ per oz.
Ⓓ 66¢ per oz.

8. If Carol pays $62.90 to fill the 17-gallon gas tank in her vehicle and she can drive 330 miles on one tank of gas, about how much does she pay per mile to drive her vehicle?

Ⓐ $0.37
Ⓑ $3.70
Ⓒ $0.19
Ⓓ $0.01

9. A 13 ounce box of cereal costs $3.99. What is the unit price per pound?

Ⓐ about $1.23
Ⓑ about $2.66
Ⓒ about $4.30
Ⓓ about $4.91

10. A bottle of perfume costs $26.00 for a $\frac{1}{2}$ ounce bottle. What is the price per ounce?

Ⓐ $25.50
Ⓑ $26.50
Ⓒ $52.00
Ⓓ $13.00

11. Check the box in each row that represents the correct unit rate for each situation.

	1:10	1:5	1:50	1:20
$1.00 per 5 pounds	○	○	○	○
50 pounds per box	○	○	○	○
10 miles per gallon	○	○	○	○
1 lap in 20 minutes	○	○	○	○

12. Below is a recipe for Grandma Grittle's favorite cupcakes. Which of the following correctly expresses a ratio found in the recipe? Check all answers that apply.

Ingredients for 12 cupcakes:
White flour - 2 cups
Sugar - 1 cup
Baking powder - 2 tsp.
Salt - 1 tsp.
Butter or margarine - 1/3 cup
Milk - 2/3 cup
Vanilla - 1 tsp.
Semi-sweet chocolate - 1 bar

Ⓐ sugar and flour, 1:2
Ⓑ flour and total cupcakes, 1:12
Ⓒ vanilla and salt, 1:1
Ⓓ salt and baking powder, 1:2
Ⓔ chocolate to total cupcakes, 1:24

13. The bag of apples shown in the picture, costs $3.20. The cost of one apple is _______.

Do NOT write your answers in this book. To open the answer sheet, scan the QR code or visit ***lumoslearning.com/a/6m003***

Chapter 1 → Lesson 3: Solving Real World Ratio Problems

1. **How many kilograms are there in 375 grams?**

   Ⓐ 3,750 kg
   Ⓑ 37.5 kg
   Ⓒ 3.75 kg
   Ⓓ 0.375 kg

2. **How many inches are there in 2 yards?**

   Ⓐ 24 in
   Ⓑ 36 in
   Ⓒ 48 in
   Ⓓ 72 in

3. **What is 50% of 120?**

   Ⓐ 50
   Ⓑ 60
   Ⓒ 70
   Ⓓ 55

4. **Michael Jordan is six feet 6 inches tall. How much is that in inches?**

   Ⓐ 66 inches
   Ⓑ 76 inches
   Ⓒ 86 inches
   Ⓓ 78 inches

5. **What is 7.5% in decimal notation?**

   Ⓐ 0.75
   Ⓑ 0.075
   Ⓒ 0.0075
   Ⓓ 7.5

6. A $60 shirt is on sale for 30% off. How much is the shirt's sale price?

Ⓐ $30
Ⓑ $40
Ⓒ $18
Ⓓ $42

7. On Monday, 6 out of every 10 people who entered a store purchased something. If 1,000 people entered the store on Monday, how many people purchased something?

Ⓐ 6 people
Ⓑ 60 people
Ⓒ 600 people
Ⓓ 610 people

8. If a pair of pants that normally sells for $51.00 is now on sale for $34.00, by what percentage was the price reduced?

Ⓐ 30%
Ⓑ 60%
Ⓒ 33.33%
Ⓓ 66.67%

9. If Comic Book World is taking 28% off the comic books that normally sell for $4.00, how much money is Kevin saving if he buys 12 comic books during the sale?

Ⓐ $28
Ⓑ $12
Ⓒ $13.44
Ⓓ $14.58

10. Eric spends 45 minutes getting to work and 45 minutes returning home. What percent of the day does Eric spend commuting?

Ⓐ 6.25%
Ⓑ 7.8%
Ⓒ 5.95%
Ⓓ 15%

11. **75% of the crowd at a sports rally were wearing the team colors. If 222 people were wearing the team colors, how many people were in the crowd? Circle the correct answer choice.**

Ⓐ 74
Ⓑ 300
Ⓒ 296
Ⓓ 314

12. **To make yummy fruit punch, use 2 cups of grape juice for every 3 cups of apple juice. Select all of the juice combinations below that correctly follow this recipe ratio.**

Ⓐ 4 cups grape juice: 6 cups apple juice
Ⓑ 5 cups grape juice: 10 cups apple juice
Ⓒ 6 cups grape juice: 9 cups apple juice
Ⓓ 6 cups grape juice: 12 cups apple juice
Ⓔ All of the above

13. **Look at the ratio information found in the table below. Complete the table by correctly filling in the missing information.**

Feet	Yards
3	1
6	
	3
15	5
24	

Do NOT write your answers in this book. To open the answer sheet, scan the QR code or visit ***lumoslearning.com/a/6m004***

Chapter 1 → Lesson 4: Solving Unit Rate Problems

1. **A 12 pack of juice pouches costs $6.00. How much does one juice pouch cost?**

   Ⓐ $0.02
   Ⓑ $0.20
   Ⓒ $0.50
   Ⓓ $0.72

2. **Eli can ride his scooter 128 miles on one tank of gas. If the scooter has a 4-gallon gas tank, how far can Eli ride on one gallon of gas?**

   Ⓐ 64 miles
   Ⓑ 32 miles
   Ⓒ 512 miles
   Ⓓ 20 miles

3. **Clifton ran 6 miles in 39 minutes. At this rate, how much time Clifton takes to run one mile?**

   Ⓐ 13 minutes
   Ⓑ 12 minutes
   Ⓒ 7.2 minutes
   Ⓓ 6 minutes and 30 seconds

4. **Brad has swimming practice 3 days a week. This week Brad swam a total of 114 laps. At this rate how many laps did Brad swim each day?**

   Ⓐ 38 laps
   Ⓑ 42 laps
   Ⓒ 57 laps
   Ⓓ 61 laps

5. Karen bought a total of seven items at five different stores. She began with $65.00 and had $15.00 remaining. Which of the following equation can be used to determine the average cost per item?

Ⓐ $7x \times 5 = \$50.00$
Ⓑ $7x = \$75.00$
Ⓒ $7x + \$15.00 = \$65.00$
Ⓓ $5x = \$65.00 - \$15.00$

6. Geoff goes to the archery range five days a week. He must pay $1.00 for every ten arrows that he shoots. If he spent $15.00 this week on arrows what is the average number of arrows Geoff shot per day?

Ⓐ 3 arrows
Ⓑ 30 arrows
Ⓒ 45 arrows
Ⓓ 75 arrows

7. Julia made 7 batches of cookies and ate 3 cookies. There were 74 cookies left. Which expression can be used to determine the average number of cookies per batch?

Ⓐ $74 \div 7$
Ⓑ $(74+7) \div 3$
Ⓒ $\frac{74 + 3}{7}$
Ⓓ $\frac{74}{3} \times 7$

8. Lars delivered 124 papers in 3 hours. How long did it take Lars to deliver one paper?

Ⓐ 1 minute
Ⓑ 1 minute and 27 seconds
Ⓒ 1 minute and 45 seconds
Ⓓ 2 minutes and 3 seconds

9. Mr. and Mrs. Fink met their son Conrad at the beach. Mr. and Mrs. Fink drove 462 miles on 21 gallons of fuel. Conrad drove 456 miles on 12 gallons of fuel. How many more miles per gallon does Conrad's car get than Mr. and Mrs. Fink's car?

Ⓐ 6 mpg
Ⓑ 22 mpg
Ⓒ 16 mpg
Ⓓ 38 mpg

10. Myka bought a box of 30 greeting cards for $4.00. Chuck bought a box of 100 greeting cards for $12.00. Who got the better deal?

Ⓐ Myka got the better deal at about 13 cents per card.
Ⓑ Myka got the better deal at about 7.5 cents per card.
Ⓒ Chuck got the better deal at about 8 cents per card.
Ⓓ Chuck got the better deal at 12 cents per card.

11. John paid $15 for 3 cheeseburgers. What is the rate of one cheeseburger? Enter your answer in the box below.

$

12. Tommy charges the same rate for each yard he mows. Calculate the rate he charges, then complete the missing information in the table.

Day	Total Money Earned $	Yards Mowed
Monday	50	2
Wednesday		3
Friday	25	
Saturday		5

13. A movie theatre charges $10 for a ticket. Check the box in each row that represents how much money the theatre would make from ticket sales.

	$150	$10	$100	$200
10 tickets	☐	☐	☐	☐
15 tickets	☐	☐	☐	☐
20 tickets	☐	☐	☐	☐
1 ticket	☐	☐	☐	☐

Do NOT write your answers in this book. To open the answer sheet, scan the QR code or visit ***lumoslearning.com/a/6m005***

Chapter 1 → Lesson 5: Finding Percent

1. **What is 25% of 24?**

Ⓐ 5
Ⓑ 6
Ⓒ 11
Ⓓ 17

2. **What is 15% of 60?**

Ⓐ 9
Ⓑ 12
Ⓒ 15
Ⓓ 25

3. **9 is what percent of 72?**

Ⓐ 7.2%
Ⓑ 8%
Ⓒ 12.5%
Ⓓ 14%

4. **How much is 30% of 190?**

Ⓐ 45
Ⓑ 57
Ⓒ 60
Ⓓ 63

5. **Daniel has 280 baseball cards. 15% of these are highly collectable. How many baseball cards does Daniel possess that are highly collectable?**

Ⓐ 15 cards
Ⓑ 19 cards
Ⓒ 42 cards
Ⓓ 47 cards

6. The football team consumed 80% of the water provided at the game. If the team consumed 8-gallons of water, how much water was provided?

Ⓐ 10 gallons
Ⓑ 12 gallons
Ⓒ 15 gallons
Ⓓ 18.75 gallons

7. Joshua brought 156 of his 678 Legos to Emily's house. What percentage of his Legos did Joshua bring?

Ⓐ 4%
Ⓑ 23%
Ⓒ 30%
Ⓓ 43%

8. At batting practice Alexis hit 8 balls out of 15 into the outfield. Which equation below can be used to determine the percentage of balls hit into the outfield?

Ⓐ $\frac{15}{8} = \frac{x}{100}$

Ⓑ $\frac{15}{100} = \frac{x}{8}$

Ⓒ $8x = (100)(15)$

Ⓓ $\frac{15}{8} = \frac{100}{x}$

9. Nikki grows roses, tulips, and carnations. She has 78 flowers of which 32% are roses. Approximately how many roses does Nikki have?

Ⓐ 18 roses
Ⓑ 25 roses
Ⓒ 28 roses
Ⓓ 41 roses

10. Victor took out 30% of his construction paper. Of this, Paul used 6 sheets, Allison used 8 sheets and Victor and Gayle used the last ten sheets. How many sheets of construction paper did Victor not take out?

Ⓐ 24 sheets
Ⓑ 50 sheets
Ⓒ 56 sheets
Ⓓ 80 sheets

11. The following items were bought on sale. Complete the missing information.

Item Purchased	Original Price	Amount of Discount	Amount Paid
Video Game	$80	20%	
Movie Ticket	$14		$11.20
Laptop	$1,000		$750
Shoes	$55.00	10%	$49.5

12. Which of these represent 25 percent of the beginning number? Select all that apply.

Ⓐ $90, $22.50
Ⓑ $150, $15.00
Ⓒ $400, $300.00
Ⓓ $560.00, $140.00

Do NOT write your answers in this book. To open the answer sheet, scan the QR code or visit ***lumoslearning.com/a/6m006***

Chapter 1 → Lesson 6: Measurement Conversion

1. Owen is 69 inches tall. How tall is Owen in feet?

Ⓐ 5.2 feet
Ⓑ 5.75 feet
Ⓒ 5.9 feet
Ⓓ 6 feet

2. What is 7 gallons 3 quarts expressed as quarts?

Ⓐ 4.75 quarts
Ⓑ 28 quarts
Ⓒ 29.2 quarts
Ⓓ 31 quarts

3. How many centimeters in 3.7 kilometers?

Ⓐ 0.000037 cm
Ⓑ 0.037 cm
Ⓒ 3700 cm
Ⓓ 370,000 cm

4. 136 ounces is how many pounds?

Ⓐ 6.8 pounds
Ⓑ 8.5 pounds
Ⓒ 1088 pounds
Ⓓ 2,176 pounds

5. How many ounces in 5 gallons?

Ⓐ 128 ounces
Ⓑ 320 ounces
Ⓒ 640 ounces
Ⓓ 1280 ounces

6. Lisa, Susan, and Chris participated in a three-person relay team. Lisa ran 1284 meters, Susan ran 1635 meters and Chris ran 1473 meters. How long was the race in kilometers? Round your answer to the nearest tenth.

Ⓐ 4.0 km
Ⓑ 4.4 km
Ⓒ 43.9 km
Ⓓ 49.0 km

7. Quita recorded the amount of time it took her to complete her chores each week for a month; 1 hour 3 minutes, 1 hour 18 minutes, 55 minutes, and 68 minutes. How many hours did Quita spend doing chores during the month?

Ⓐ 3.8 hours
Ⓑ 4.24 hours
Ⓒ 4.4 hours
Ⓓ 5.7 hours

8. Lamar can run 3 miles in 18 minutes. At this rate, how much distance he can run in one hour?

Ⓐ 0.9 miles
Ⓑ 1.1 miles
Ⓒ 10 miles
Ⓓ 21 miles

9. A rectangular garden has a width of 67 inches and a length of 92 inches. What is the perimeter of the garden in feet?

Ⓐ 13.25 feet
Ⓑ 26.5 feet
Ⓒ 31.8 feet
Ⓓ 42.8 feet

10. Pat has a pen pal in England. When Pat asked how tall his pen pal was he replied, 1.27 meters. If 1 inch is 2.54 cm, how tall is Pat's pen pal in feet and inches?

Ⓐ 3 feet 11 inches
Ⓑ 4 feet 2 inches
Ⓒ 4 feet 6 inches
Ⓓ 5 feet exactly

11. How many fluid ounces are there in a cup? Circle the correct answer choice.

Ⓐ 10 fl oz.
Ⓑ 8 fl oz.
Ⓒ 4 fl oz.
Ⓓ 16 fl oz.

12. How many meters are there in 16 kilometers? Circle the correct answer choice.

Ⓐ 1.6 m
Ⓑ 1,600 m
Ⓒ 16,000 m
Ⓓ 160 m

13. Use the chart provided to fill in the missing values in the table below.

1 L	1000 ml
1 g	1000 mg
1 m	1000 mm

3	L		ml
	g	5000	mg
	m	8000	mm
12	L		ml
20	g		mg

End of Ratios & Proportional Relationships

Chapter 2
The Number System

Do NOT write your answers in this book. To open the answer sheet, scan the QR code or visit ***lumoslearning.com/a/6m007***

Lesson 1: Division of Fractions

1. What is the quotient of 20 divided by one-fourth?

Ⓐ 80
Ⓑ 24
Ⓒ 5
Ⓓ 15

2. Calculate: $1\frac{1}{2} \div \frac{3}{4} =$

Ⓐ 4

Ⓑ $\frac{1}{2}$

Ⓒ $\frac{3}{4}$

Ⓓ 2

3. Calculate: $3\frac{2}{3} \div 2\frac{1}{6} =$

Ⓐ $\frac{8}{13}$

Ⓑ $\frac{12}{13}$

Ⓒ $1\frac{5}{13}$

Ⓓ $1\frac{9}{13}$

4. **Calculate:** $2\frac{3}{4} \div \frac{11}{4} =$

Ⓐ 1
Ⓑ 2
Ⓒ 3
Ⓓ 4

5. **Calculate:** $\frac{7}{8} \div \frac{3}{4} =$

Ⓐ $1\frac{1}{6}$

Ⓑ 2

Ⓒ $\frac{21}{32}$

Ⓓ $\frac{5}{9}$

6. **Calculate:** $6\frac{3}{4} \div 1\frac{1}{8} =$

Ⓐ $\frac{1}{6}$

Ⓑ 4

Ⓒ $5\frac{3}{4}$

Ⓓ 6

7. **Complete the following division using mental math.**

7 divided by $\frac{1}{5}$

Ⓐ 35

Ⓑ $\frac{7}{5}$

Ⓒ $\frac{5}{7}$

Ⓓ $\frac{1}{35}$

8. **Complete the following division using mental math.**

11 divided by $\frac{6}{6}$

Ⓐ $\frac{66}{66}$
Ⓑ $\frac{1}{11}$
Ⓒ 1
Ⓓ 11

9. **What is the result when a fraction is multiplied by its reciprocal?**

Ⓐ $\frac{1}{2}$
Ⓑ 10
Ⓒ 1
Ⓓ It cannot be determined.

10. **Simplify the following problem. Do not solve.**

$\frac{14}{21} \div \frac{28}{7}$

Ⓐ $\frac{14}{21} \div \frac{28}{7}$
Ⓑ $\frac{2}{3} \times \frac{1}{4}$
Ⓒ 1
Ⓓ 10

11. Which of the following is equal to $1 \div \frac{3}{4}$? Circle the correct answer choice.

Ⓐ $\frac{4}{3}$

Ⓑ $\frac{2}{4}$

Ⓒ $\frac{1}{3}$

12. Fill in the blank.

$\frac{1}{2} \div 4 =$ ___?

13. Which of the following is equal to $\frac{7}{2} \div \frac{2}{6}$? Circle the correct answer choice.

Ⓐ $\frac{9}{2}$

Ⓑ $\frac{5}{4}$

Ⓒ $\frac{42}{4}$

Do NOT write your answers in this book. To open the answer sheet, scan the QR code or visit ***lumoslearning.com/a/6m008***

Chapter 2 → Lesson 2: Division of Whole Numbers

1. **A team of 12 players got an award of $1,800 for winning a championship football game. If the captain of the team is allowed to keep $315, how much money would each of the other players get? (Assume they split it equally.)**

   Ⓐ $135
   Ⓑ $125
   Ⓒ $150
   Ⓓ $123.75

2. **Peter gets a salary of $125 per week. He wants to buy a new television that costs $3,960. If he saves $55 per week, which of the following expressions could he use to figure out how many weeks it will take him to save up enough money to buy the new TV?**

   Ⓐ $3,960 ÷ ($125 – $55)
   Ⓑ $3,960 – ($125)($55)
   Ⓒ ($3,960 ÷ $125) ÷ $55
   Ⓓ $3,960 ÷ $55

3. **An expert typist typed 9,000 words in two hours. How many words per minute did she type?**

   Ⓐ 4,500 words per minute
   Ⓑ 150 words per minute
   Ⓒ 75 words per minute
   Ⓓ 38 words per minute

4. **Bethany cut off 18 inches of her hair for "Locks of Love". (Locks of Love is a non profit organization that provides wigs to people who have lost their hair due to chemotherapy.) It took her 3 years to grow it back. How much did her hair grow each month?**

   Ⓐ 1 inch
   Ⓑ 2 inches
   Ⓒ 0.25 inches
   Ⓓ 0.5 inches

5. On "Jeopardy," during the month of September, the champions won a total of $694,562. Assuming that there were 22 "Jeopardy" shows in September, what was the average amount won each day by the champions?

Ⓐ $12,435
Ⓑ $21,891
Ⓒ $35,176
Ⓓ $31,571

6. A marching band wants to raise $20,000 at its annual fundraiser. If they sell tickets for $20 a piece, how many tickets will they have to sell?

Ⓐ 500
Ⓑ 10,000
Ⓒ 100
Ⓓ 1,000

7. A classroom needs 3,200 paper clips for a project. If there are 200 paper clips in a package, how many packages will they need in all?

Ⓐ 160
Ⓑ 1,600
Ⓒ 18
Ⓓ 16

8. A homebuilder is putting new shelves in each closet he is building. He has 2,592 shelves in his inventory. If each closet needs 108 shelves, how many closets can he build?

Ⓐ 2.4
Ⓑ 108
Ⓒ 42
Ⓓ 24

9. A toy maker needs to make $17,235 per month to meet his costs. Each toy sells for $45. How many toys does he need to sell in order to break even (cover his costs)?

Ⓐ 393
Ⓑ 473
Ⓒ 373
Ⓓ 383

10. A stamp collector collected 4,224 stamps last year. He collected the same amount each month. How many stamps did he collect each month?

Ⓐ 422
Ⓑ 352
Ⓒ 362
Ⓓ 252

11. Fill in the blank

40,950 ÷ ____ = 26

12. Fill in the blank

455 ÷ ____ = 7

Do NOT write your answers in this book. To open the answer sheet, scan the QR code or visit ***lumoslearning.com/a/6m009***

Chapter 2 → Lesson 3: Operations with Decimals

1. **Three friends went out to lunch together. Ben got a meal that cost $7.25, Frank got a meal that cost $8.16, and Herman got a meal that cost $5.44. If they split the check evenly, how much did they each pay for lunch? (Assume no tax)**

   Ⓐ $6.95
   Ⓑ $7.75
   Ⓒ $7.15
   Ⓓ $6.55

2. **Which of these is the standard form of twenty and sixty-three thousandths?**

   Ⓐ 20.63000
   Ⓑ 20.0063
   Ⓒ 20.63
   Ⓓ 20.063

3. **Mr. Zito bought a bicycle for $160. He spent $12.50 on repair charges. If he sold the same bicycle for $215, what would his profit be on the investment?**

   Ⓐ $ 147.50
   Ⓑ $ 42.50
   Ⓒ $ 67.50
   Ⓓ $ 55.00

4. **A certain book is sold in a paperback version for $4.75 or in a hardcover version for $11.50. If a copy of the book is being purchased for each of the twenty students in Mrs. Jackson's class, how much money altogether would be saved by buying the paperback version, as opposed to the hardcover version?**

   Ⓐ $ 155.00
   Ⓑ $ 135.00
   Ⓒ $ 115.00
   Ⓓ $ 145.00

5. Which of these sets contains all equivalent numbers?

Ⓐ $\left\{0.75, \frac{3}{4}, 75\%, \frac{8}{12}\right\}$

Ⓑ $\left\{0.100, \frac{5}{50}, 15\%, 0.010\right\}$

Ⓒ $\left\{\frac{3}{8}, 35\%, 0.35, \frac{35}{100}\right\}$

Ⓓ $\left\{\frac{9}{25}, 36\%, 0.360, \frac{18}{50}\right\}$

6. Brian is mowing his lawn. He and his family have 7.84 acres. Brian mows 1.29 acres on Monday, 0.85 acres on Tuesday, and 3.63 acres on Thursday. How many acres does Brian have left to mow?

Ⓐ 2.70
Ⓑ 20.7
Ⓒ 2.07
Ⓓ 0.207

7. Hector is planting his garden. He makes it 5.8 feet wide and 17.2 feet long. What is the area of Hector's garden?

Ⓐ 9.976 square feet
Ⓑ 99.76 square feet
Ⓒ 99.76 feet
Ⓓ 997.6 square feet

8. Chris and 2 of his friends go apple picking. Together they pick a bushel of apples that weighs 28.2 pounds. If Chris and his friends split the bushel of apples evenly among themselves, how many pounds of apples will each person take home?

Ⓐ 9.4 pounds
Ⓑ 0.94 pounds
Ⓒ 94 pounds
Ⓓ 0.094 pounds

9. Joann and John are hiking over a three-day weekend. They have a total of 67.8 miles that they are planning on hiking. On Friday, they hike half of the miles. On Saturday, they hike another 20 miles. How many miles do they have left to hike on Sunday?

Ⓐ 1.39 miles
Ⓑ 31.9 miles
Ⓒ 13.9 miles
Ⓓ 139 miles

10. Margaret, Justin, and Leigh are babysitting for the neighbor's children during the summer. Each week they make a total of $72.00, and they split the money evenly. At the end of 4 weeks, how much money did Justin make?

Ⓐ $130.00
Ⓑ $144.00
Ⓒ $72.00
Ⓓ $96.00

11. Match the equation with the correct answer.

	18.711	1871.1	187.11
62.37 x 30 =			
6.237 x 30 =			
0.6237 x 30 =			

12. Fill in the blank.

7.1 × 3.2 = ____

13. Match the equation with the correct answer.

	6.67	88.84	910.54
9.13 – 2.46 =			
913 – 2.46 =			
91.3 – 2.46 =			

Do NOT write your answers in this book. To open the answer sheet, scan the QR code or visit ***lumoslearning.com/a/6m010***

Chapter 2 → Lesson 4: Using Common Factors

1. Which of these statements is true of the number 17?

Ⓐ It is a factor of 17.
Ⓑ It is a multiple of 17.
Ⓒ It is prime.
Ⓓ All of the above are true.

2. What are the single digit prime numbers?

Ⓐ 2, 3, 5, and 7
Ⓑ 1, 2, 3, 5, and 7
Ⓒ 3, 5, and 7
Ⓓ 1, 3, 5, and 7

3. Which of the following sets below contains only prime numbers?

Ⓐ 7, 11, 49
Ⓑ 7, 37, 51
Ⓒ 7, 23, 47
Ⓓ 2, 29, 93

4. The product of three numbers is equal to 105. If the first two numbers are 7 and 5, what is the third number?

Ⓐ 35
Ⓑ 7
Ⓒ 5
Ⓓ 3

5. What is the prime factorization of 240?

Ⓐ $2 \times 2 \times 2 \times 5$
Ⓑ $2 \times 2 \times 2 \times 2 \times 15$
Ⓒ $2 \times 2 \times 2 \times 2 \times 5 \times 3$
Ⓓ $2 \times 2 \times 2 \times 5 \times 3$

6. Office A's building complex and the school building next door have the same number of rooms. Office A's building complex has floors with 5 one-room offices on each, and the school building has 11 classrooms on each floor. What is the fewest number of rooms that each building can have?

Ⓐ 16
Ⓑ 6
Ⓒ 55
Ⓓ $\frac{5}{11}$

7. How do you know if a number is divisible by 3?

Ⓐ if the ones digit is an even number
Ⓑ if the ones digit is 0 or 5
Ⓒ if the sum of the digits in the number is divisible by 3 or a multiple of 3
Ⓓ if the sum of the digits in the number is divisible by 2 and 3

8. Find the Greatest Common Factor (GCF) for 42 and 56.

Ⓐ 21
Ⓑ 14
Ⓒ 7
Ⓓ 2

9. What is the prime factorization of 110?

Ⓐ 10 × 11
Ⓑ 110 × 1
Ⓒ 55 × 2
Ⓓ 2 × 5 × 11

10. What is the LCM of 16 and 24?

Ⓐ 16
Ⓑ 24
Ⓒ 36
Ⓓ 48

11. Fill in the blank.

The Greatest Common Factor (GCF) of 24, 36, and 48 is ________.

12. List all the factors of 20.

13. Circle the common factors between 10 and 15.

Ⓐ 1, 5
Ⓑ 1, 2, 5
Ⓒ 1, 3, 4

14. What are the common factors of 8 and 12? Write your answer in the box below.

Do NOT write your answers in this book. To open the answer sheet, scan the QR code or visit ***lumoslearning.com/a/6m011***

Chapter 2 → Lesson 5: Positive and Negative Numbers

1. Larissa has $4\frac{1}{2}$ cups of flour. She is making cookies using a recipe that calls for $2\frac{3}{4}$ cups of flour. After baking the cookies how much flour will be left?

Ⓐ $2\frac{3}{4}$ cups

Ⓑ $2\frac{1}{4}$ cups

Ⓒ $2\frac{3}{8}$ cups

Ⓓ $1\frac{3}{4}$ cups

2. The accounting ledger for the high school band showed a balance of $2,123. They purchased new uniforms for a total of $2,400. How much must they deposit into their account in order to prevent it from being overdrawn?

Ⓐ $382
Ⓑ $462
Ⓒ $4,000
Ⓓ $277

3. Juan is climbing a ladder. He begins on the first rung, climbs up four rungs, but then slides down two rungs. What rung is Juan on?

Ⓐ 2
Ⓑ 3
Ⓒ 4
Ⓓ 5

4. **If last year Julie's net profit was $26,247 after she spent $14,256 on expenses, what was her gross revenue?**

Ⓐ −$40,503
Ⓑ −$11,991
Ⓒ $40, 503
Ⓓ $11,991

5. **The amount of snow on the ground increased by 4 inches between 4 p.m. and 6 p.m. If there was 6 inches of snow on the ground at 4 p.m. how many inches were on the ground at 6 p.m.?**

Ⓐ 10 inches
Ⓑ 14 inches
Ⓒ 2 inches
Ⓓ 18 inches

6. **The temperature at noon was 20 degrees F. For the next 5 hours it dropped 2 degrees per hour. What was the temperature at 5:00 p.m.?**

Ⓐ 15 degrees F
Ⓑ 10 degrees F
Ⓒ 5 degrees F
Ⓓ 0 degrees F

7. **Tom enters an elevator that is in the basement, one floor below ground level. He travels three floors down to the parking level and then 4 floors back up. What floor does he end up on?**

Ⓐ ground level
Ⓑ 2nd floor
Ⓒ 3rd floor
Ⓓ basement

8. **Which of these numbers would not be found between 6 and 7 on a number line?**

Ⓐ $\frac{43}{6}$

Ⓑ $\frac{34}{5}$

Ⓒ $\frac{19}{3}$

Ⓓ $\frac{100}{16}$

9. Stacey lives on a cliff. She lives 652 feet above sea level. When she travels to town, she travels down 491 feet. What elevation is town?

Ⓐ 261 feet above sea level
Ⓑ 161 feet below sea level
Ⓒ 161 feet above sea level
Ⓓ 141 feet above sea level

10. Chris lives in Alaska. When he wakes up, the temperature is −53 degrees F. By noon, the temperature has risen 27 degrees F. What is the temperature at noon?

Ⓐ −27 degrees F
Ⓑ −26 degrees F
Ⓒ 26 degrees F
Ⓓ −24 degrees F

11. Which of the following would result in a negative answer? Choose all that apply.

Ⓐ 25 - 56
Ⓑ 15 + 42
Ⓒ -15 + 8
Ⓓ -25 – 7
Ⓔ 74 – 32

12. On a number line, how far apart are -27 and 30? Write your answer in the box below.

13. Which among the following has the lowest value?

-15, 0, 10, 25

Enter your answer in the box below.

Do NOT write your answers in this book. To open the answer sheet, scan the QR code or visit ***lumoslearning.com/a/6m012***

Chapter 2 → Lesson 6: Representing Negative Numbers

1. Which of these numbers would be found closest to 0 on a number line?

Ⓐ −5

Ⓑ $-5\frac{1}{2}$

Ⓒ $4\frac{1}{2}$

Ⓓ −4

2. On a number line, how far apart are the numbers −5.5 and 7.5?

Ⓐ 13 units
Ⓑ 12 units
Ⓒ 12.5 units
Ⓓ 2 units

3. Which numbers does the following number line represent?

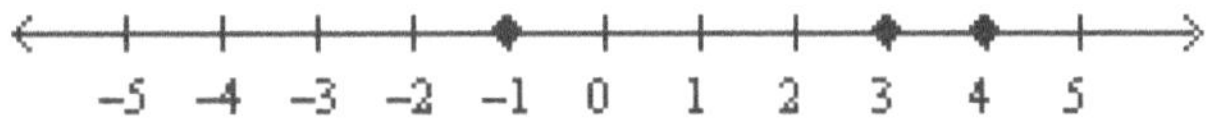

Ⓐ {−2, 0, 5}
Ⓑ {−3, −1, 4}
Ⓒ {−1, 3, 5}
Ⓓ {−1, 3, 4}

4. If you go left on a number line, you will ____________________

Ⓐ go in a negative direction
Ⓑ go in a positive direction
Ⓒ increase your value
Ⓓ none of these

5. Which of the following numbers is NOT between −7 and −4 on the number line?

Ⓐ −7.2
Ⓑ −4.8
Ⓒ −6
Ⓓ −5.01

6. −7.25 is between which two numbers on the number line?

Ⓐ −7 and −6
Ⓑ −5 and −3
Ⓒ −9 and −10
Ⓓ −7 and −8

7. What happens when you start at any number on a number line and add its additive inverse?

Ⓐ The number doubles.
Ⓑ The number halves.
Ⓒ The sum is zero.
Ⓓ There is no movement.

8. Which set of numbers would be found to the left of 4 on the number line?

Ⓐ {−1, 4, −5}
Ⓑ {1, −4, −5}
Ⓒ {1, 4, −5}
Ⓓ {1, 4, 5}

9. On a number line, Bill places a marble on 12. He rolls the marble to the left and it moves 21 spaces and then rolls back to the right 3 spaces. What number does the marble end up on?

Ⓐ −21
Ⓑ −8
Ⓒ −9
Ⓓ −6

10. Which number would be found on the number line between 0 and −1?

Ⓐ −1.2
Ⓑ 1.2
Ⓒ −0.8
Ⓓ 0.8

11. Circle the number with the highest value.

Ⓐ −17
Ⓑ −28
Ⓒ −36

12. Which number(s) have a value lower than 3? Select all that apply.

Ⓐ −5
Ⓑ 6
Ⓒ −18
Ⓓ −23
Ⓔ 4

Do NOT write your answers in this book. To open the answer sheet, scan the QR code or visit ***lumoslearning.com/a/6m013***

Chapter 2 → Lesson 7: Ordered Pairs

1. In what Quadrant (I, II, III, IV) does the point (−12, 20) lie?

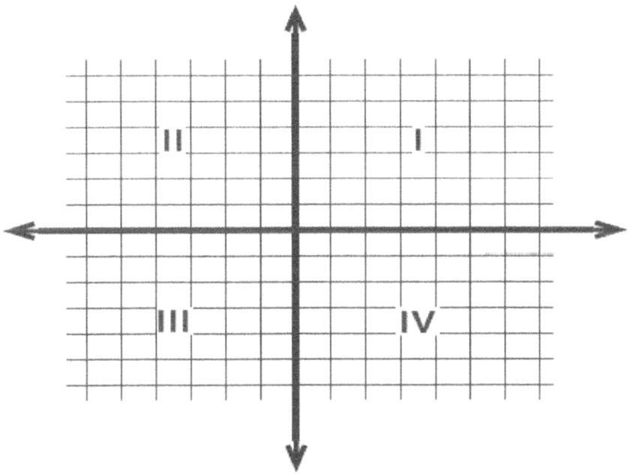

Ⓐ Quadrant I
Ⓑ Quadrant II
Ⓒ Quadrant III
Ⓓ Quadrant IV

2. In what Quadrant (I, II, III, IV) does the point (8, −9) lie?

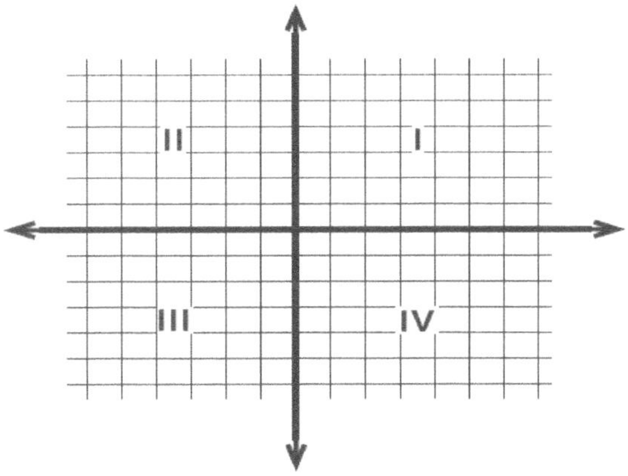

Ⓐ Quadrant I
Ⓑ Quadrant II
Ⓒ Quadrant III
Ⓓ Quadrant IV

3. **In what Quadrant (I, II, III, IV) does the point (−0.75, −0.25) lie?**

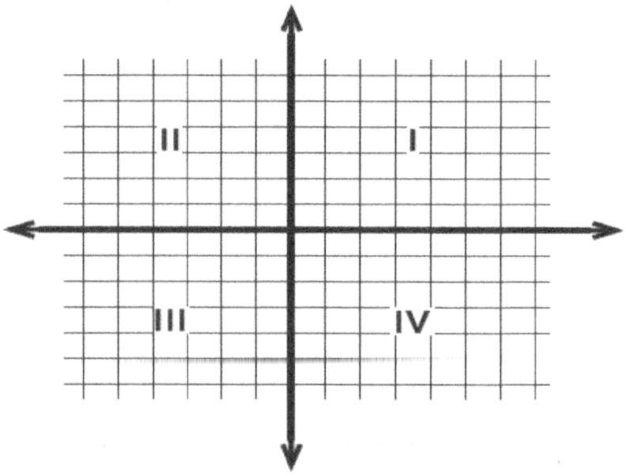

Ⓐ Quadrant I
Ⓑ Quadrant II
Ⓒ Quadrant III
Ⓓ Quadrant IV

4. **Point A, located at (10,−4), is reflected across the x-axis. What are the coordinates of the reflected point?**

Ⓐ (−10, −4)
Ⓑ (10, −4)
Ⓒ (10, 4)
Ⓓ (−10, 4)

5. **Point B, located at (6,3), is reflected across the y-axis. What are the coordinates of the reflected point?**

Ⓐ (6, 3)
Ⓑ (−6, −3)
Ⓒ (6, −3)
Ⓓ (−6, 3)

6. **Point C, located at (2, 9), is reflected across the x-axis called Point C'. Point C' is then reflected across the y-axis called Point C". What are the coordinates of Point C"?**

Ⓐ (−2, −9)
Ⓑ (2, 9)
Ⓒ (2, −9)
Ⓓ (−2, 9)

7. **Which Quadrant contains the most points in the following list of ordered pairs? (3, 4), (−2,4), (3, −2), (4, −3), (−9, −5), (1, −3), (4, −12), (6, 14), (−5, −11), (−99, −43)**

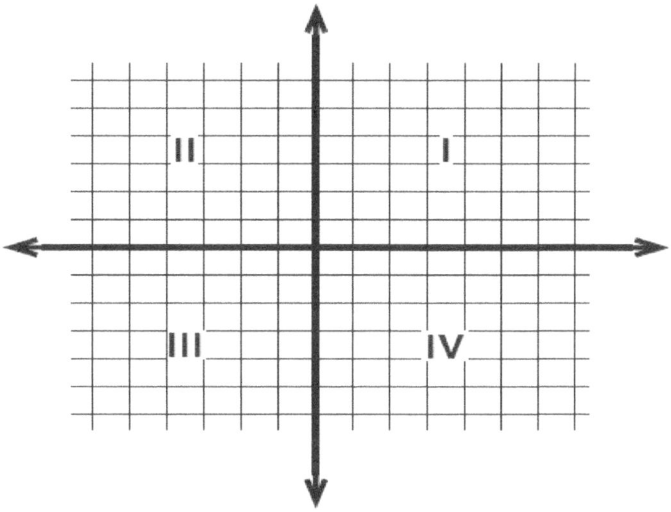

Ⓐ Quadrant I
Ⓑ Quadrant II
Ⓒ Quadrant III
Ⓓ Quadrant IV

8. **The following points have been reflected across the y-axis: (5, 1), (1, −3), (−5, 3), (−3, 1), (6, −2), (−8, 4), (7, 12). How many of the reflected points fall in Quadrant II?**

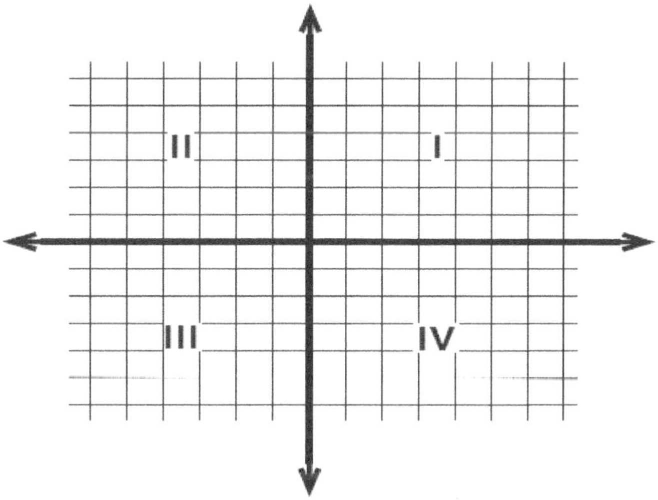

Ⓐ 0
Ⓑ 1
Ⓒ 2
Ⓓ 3

9. The following four points were reflected across the y-axis. A (4, 0), B (0, 0), C (3,0), D(–6,0). Which of the four reflected points (A', B', C', D') is not graphed properly below?

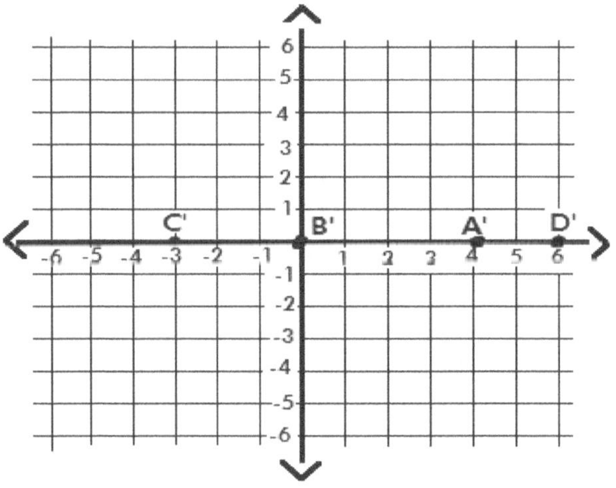

Ⓐ Point A'
Ⓑ Point B'
Ⓒ Point C'
Ⓓ Point D'

10. A point located at (12, –4) is reflected across the x-axis. In which quadrant (I, II, III, IV) will the reflected point be located?

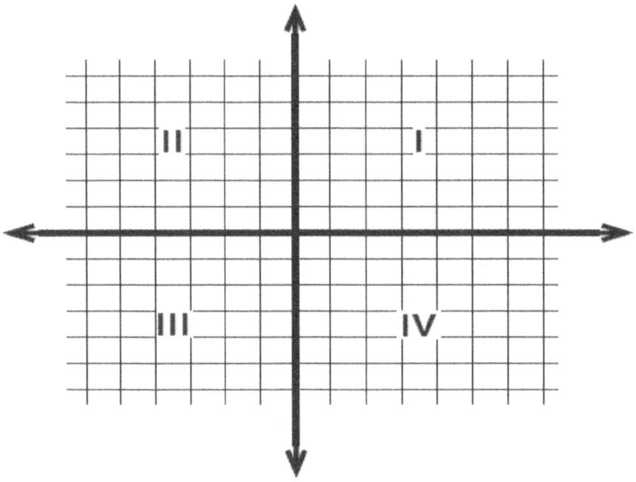

Ⓐ Quadrant I
Ⓑ Quadrant II
Ⓒ Quadrant III
Ⓓ Quadrant IV

11. Circle the correct ordered pair for the point plotted below.

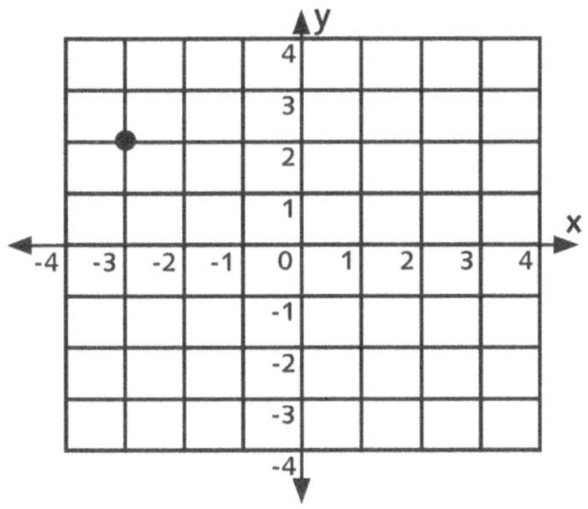

Ⓐ (3,2)
Ⓑ (-3,-2)
Ⓒ (-3,2)

12. What are the coordinates of point 'A' ? Enter your answer in the box below.

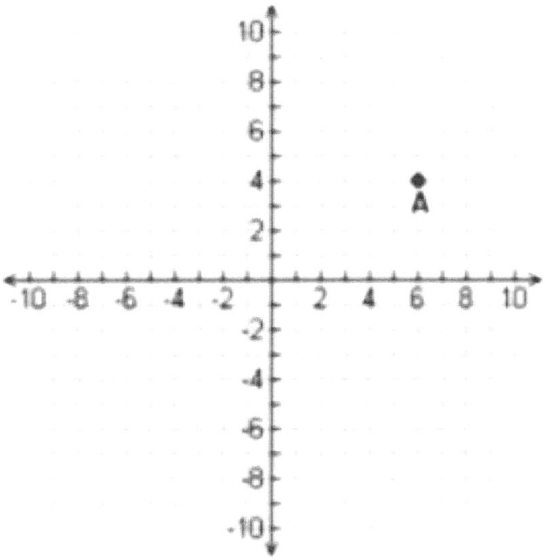

Do NOT write your answers in this book. To open the answer sheet, scan the QR code or visit ***lumoslearning.com/a/6m014***

Chapter 2 → Lesson 8: Number Line & Coordinate Plane

1. What number does the dot represent on the number line?

Ⓐ 12
Ⓑ 13
Ⓒ 14
Ⓓ 15

2. What number does the dot represent on the number line?

Ⓐ −2
Ⓑ −1
Ⓒ 0
Ⓓ 1

3. What number does the dot represent on the number line?

Ⓐ 0
Ⓑ 1
Ⓒ 4
Ⓓ 6

4. **What number does the dot represent on the number line?**

Ⓐ −5
Ⓑ −7.5
Ⓒ −8
Ⓓ −8.5

5. **What number does the dot represent on the number line?**

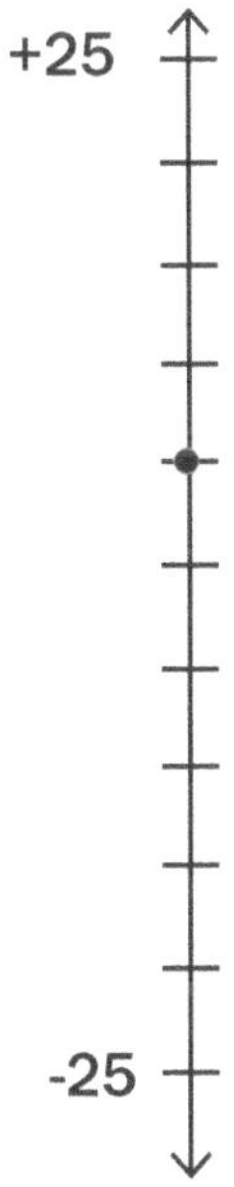

Ⓐ −5
Ⓑ 0
Ⓒ 5
Ⓓ 10

6. **Select the point located at (1,−2)**

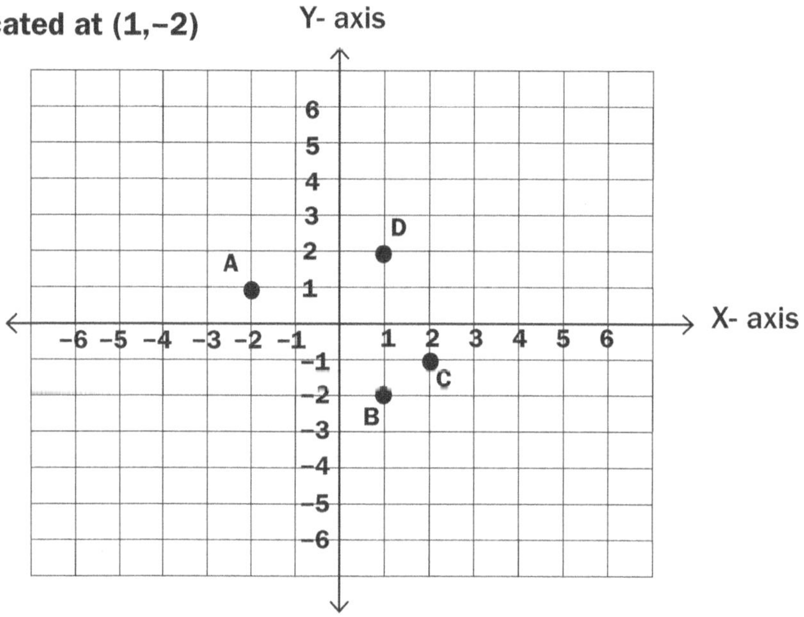

Ⓐ Point A
Ⓑ Point B
Ⓒ Point C
Ⓓ Point D

7. **Select the point located at (−3,5)**

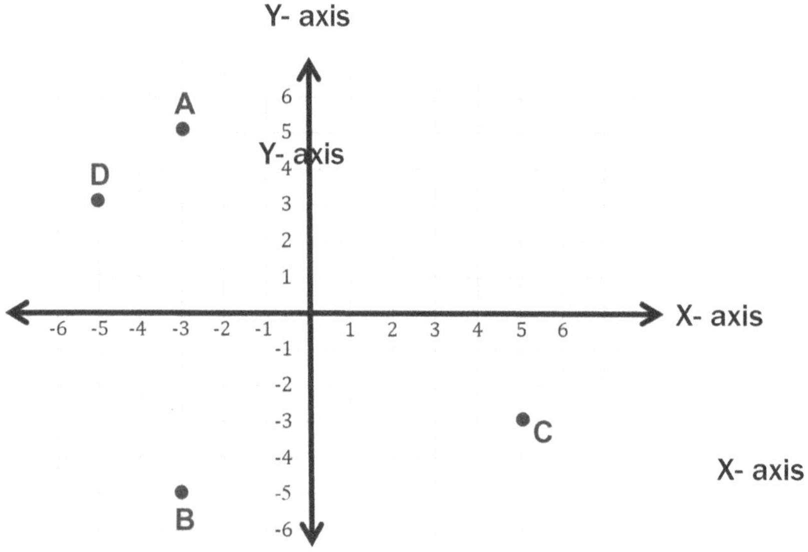

Ⓐ Point A
Ⓑ Point B
Ⓒ Point C
Ⓓ Point D

8. **Select the point located at (−4,−5)**

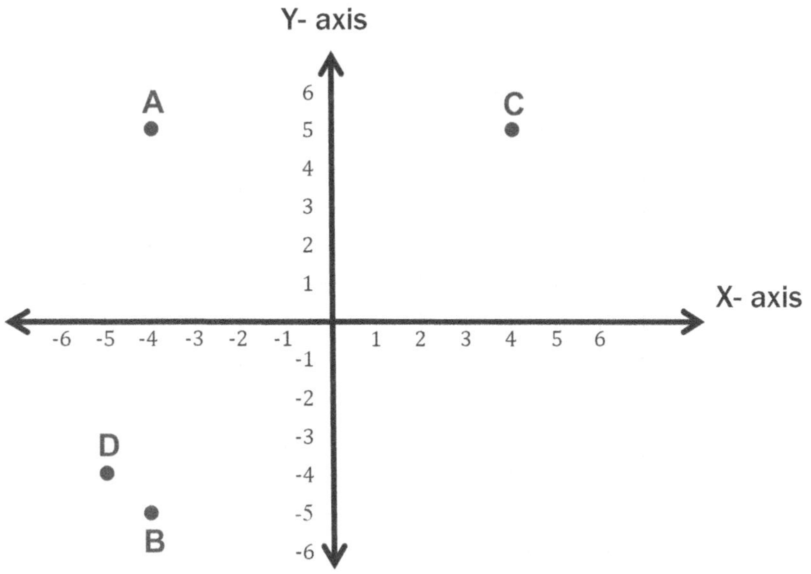

Ⓐ Point A
Ⓑ Point B
Ⓒ Point C
Ⓓ Point D

9. **Select the point located at (−1.5, 0.5)**

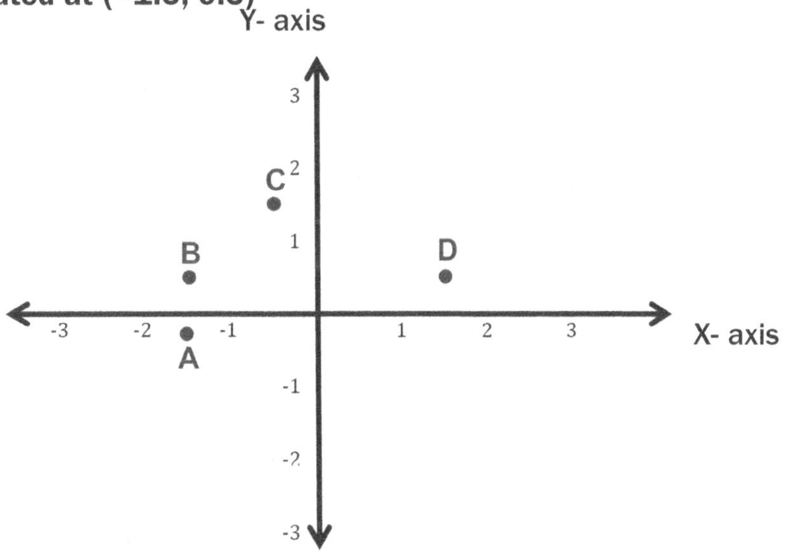

Ⓐ Point A
Ⓑ Point B
Ⓒ Point C
Ⓓ Point D

10. **Which of the following points is not graphed below? (3, 4), (−2,4), (3, 2), (−4,−3)**

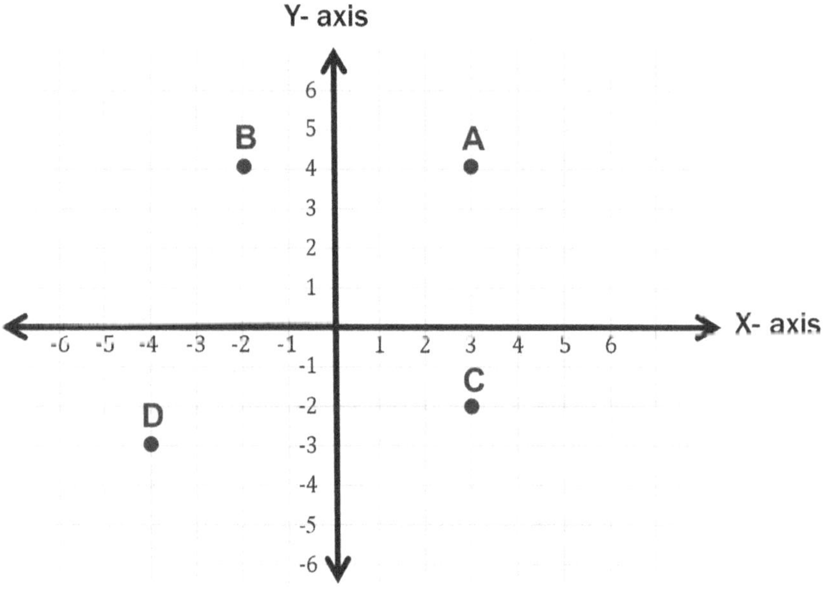

Ⓐ (3, 4)
Ⓑ (−2, 4)
Ⓒ (3, 2)
Ⓓ (−4, −3)

11. **Circle the point that names the ordered pair (-9, -2).**

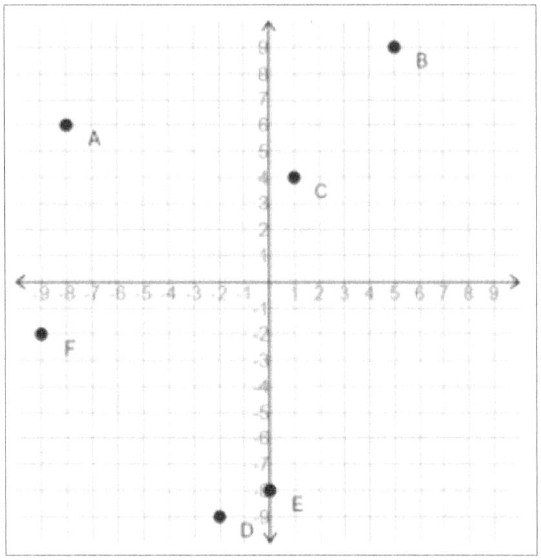

Point A	Point B	Point C
Point D	Point E	Point F

12. Choose the set of numbers that are correctly ordered from least to greatest. Use the number line to help you. Choose all answers that apply.

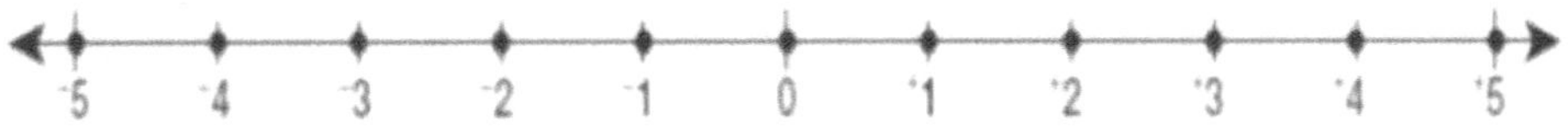

Ⓐ 5, 2, 0, -1, -3
Ⓑ -3, -1, 0, 3, 5
Ⓒ -5, -2, 1, 4, 5
Ⓓ 0, 1, 3, -4, -5
Ⓓ None of the above

Do NOT write your answers in this book. To open the answer sheet, scan the QR code or visit ***lumoslearning.com/a/6m015***

Chapter 2 → Lesson 9: Absolute Value

1. **Evaluate the following: 17 – |(7)(–3)|**

   Ⓐ 38
   Ⓑ –4
   Ⓒ 4
   Ⓓ 13

2. **Evaluate the following: 16 + |(7)(–3) – 44| – 5**

   Ⓐ 76
   Ⓑ 86
   Ⓒ 34
   Ⓓ 44

3. **Evaluate the following: |15 – 47| + 9 – |(–2)(–4) – 17|**

   Ⓐ 32
   Ⓑ –32
   Ⓒ 50
   Ⓓ 76

4. **Evaluate the following: 18 + 3 |6 – 25| – 11**

   Ⓐ 64
   Ⓑ 100
   Ⓒ 122
   Ⓓ 57

5. **Evaluate the following: 77 – |(–8)(3) + (–10)(4)|**

   Ⓐ 13
   Ⓑ 141
   Ⓒ –141
   Ⓓ –93

6. **Evaluate the following: 21 – |8 – (–5)(7)| – 54**

Ⓐ –6
Ⓑ 10
Ⓒ –118
Ⓓ –76

7. **Evaluate: |9 – 3k| = 3**

Ⓐ k = 0
Ⓑ k = 2
Ⓒ k = 2 or k = 4
Ⓓ k = –4

8. **Is the absolute value of a negative integer positive or negative?**

Ⓐ Positive
Ⓑ Negative
Ⓒ Neither
Ⓓ It depends on the magnitude of the number.

9. **What is the value of –|24|?**

Ⓐ 1/24
Ⓑ –1/24
Ⓒ –24
Ⓓ 24

10. **What is the value of |45 – 75|?**

Ⓐ –115
Ⓑ 115
Ⓒ 30
Ⓓ –30

11. **What is the value of |14| – |–28|? Write your answer in the box below.**

12. **Which is greater, |24| or |–25|? Write the answer in the box below.**

Do NOT write your answers in this book. To open the answer sheet, scan the QR code or visit ***lumoslearning.com/a/6m016***

Chapter 2 → Lesson 10: Rational Numbers in Context

1. **Xavier has a golf score of −7 (We write −7 because it is 7 points below par) and Curtis has a golf score of −12. Who has the higher score?**

   Ⓐ Xavier has the higher score.
   Ⓑ Curtis has the higher score.
   Ⓒ Because both scores are below par, neither one has the higher score.
   Ⓓ We cannot tell who has the higher score because we do not know what par is.

2. **Kelly has read $\frac{5}{6}$ of a book. Helen has read $\frac{9}{12}$ of the same book. Who has read more of the book?**

   Ⓐ $\frac{5}{6}$ is more than $\frac{9}{12}$ so Kelly has read more.
   Ⓑ $\frac{5}{6}$ is less than $\frac{9}{12}$ so Helen has read more.
   Ⓒ $\frac{5}{6}$ is the same as $\frac{9}{12}$ so both Kelly and Helen have read the same amount.
   Ⓓ We cannot tell who has read more because the fractions have different denominators.

3. **The record low temperature for NY is −52°F. The record low temperature for Alaska is −80°F. Which of the following inequalities accurately compares these two temperatures?**

   Ⓐ $-52°\text{ F} < -80°\text{ F}$
   Ⓑ $-80°\text{ F} > -52°\text{ F}$
   Ⓒ $-52°\text{ F} = -80°\text{ F}$
   Ⓓ $-52°\text{ F} > -80°\text{ F}$

4. **A cake recipe calls for $1\frac{3}{4}$ cups of soy flour, $\frac{20}{8}$ cups of rice flour and 1.6 cups of wheat flour. Which of the following inequalities compares these three quantities accurately?**

Ⓐ $1.6 < \frac{20}{8} < 1\frac{3}{4}$
Ⓑ $\frac{20}{8} > 1\frac{3}{4} > 1.6$
Ⓒ $\frac{20}{8} > 1\frac{3}{4} < 1.6$
Ⓓ $1\frac{3}{4} > 1.6 > \frac{20}{8}$

5. **Doug and Sissy went scuba diving. Doug descended to −143 feet and Sissy descended to −134 feet. Who dove deeper?**

Ⓐ −134 > −143, so Sissy dove deeper.
Ⓑ −134 < −143, so Doug dove deeper.
Ⓒ −134 = −143, so neither one dove deeper as they descended the same amount.
Ⓓ |−143|>|−134|, so Doug dove deeper.

6. **At the annual town festival $\frac{4}{15}$ of the vendors sold outdoor items, 0.4 sold clothing or indoor items and $\frac{1}{3}$ sold food. Which of the following inequalities compares these three quantities accurately?**

Ⓐ $0.4 < \frac{4}{15} < \frac{1}{3}$
Ⓑ $\frac{4}{15} > 0.4 > \frac{1}{3}$
Ⓒ $0.4 > \frac{1}{3} > \frac{4}{15}$
Ⓓ $\frac{1}{3} < \frac{4}{15} < 0.4$

7. A school of fish (S1) is spotted in the ocean at 15 feet below sea level. A second school (S2) of fish is spotted at $\frac{33}{3}$ feet below sea level. A third school (S3) of fish is spotted 11.5 feet below sea level. Order these numbers from deepest to shallowest. Note: The symbol > means deeper and < means shallower

Ⓐ **S**1 < **S**3 < **S**2
Ⓑ **S**3 < **S**2 < **S**1
Ⓒ **S**1 > **S**3 > **S**2
Ⓓ **S**3 < **S**1 < **S**2

8. During the first snowfall of the year, Henderson recorded the snow fall each day. The first day $\frac{4}{5}$ of a foot fell. On the second day $\frac{5}{7}$ of a foot fell. On which day did more snow fall?

Ⓐ $\frac{4}{5} > \frac{5}{7}$, so more snow fell on the first day.
Ⓑ $\frac{4}{5} < \frac{5}{7}$, so more snow fell on the second day.
Ⓒ $\frac{4}{5} = \frac{5}{7}$, so the same amount of snow fell on both days.
Ⓓ We cannot tell from this information because the fractions have different denominators.

9. Molly has $365 in her savings account. She withdraws $415. Bill has a savings account balance of –$45. Which of the following statements is correct?

Ⓐ Both Molly and Bill owe the bank the same amount.
Ⓑ Molly owes the bank more than Bill.
Ⓒ Bill owes the bank more than Molly.
Ⓓ Neither Molly or Bill owe the bank any money.

10. Jeremiah and Farley each bought 5 boxes of energy bars. Within a week Jeremiah eats $2\frac{5}{6}$ boxes and Farley eats $1\frac{15}{9}$ boxes. Who has more left?

Ⓐ Jeremiah has more left.
Ⓑ Farley has more left.
Ⓒ They both have the same amount left.
Ⓓ We cannot tell from this information because the mixed numbers have different denominators.

11. Which temperature is hotter, 32° F or 56° F? Enter your answer in the box below.

12. Select all numbers with a value greater than -5.

Ⓐ -17
Ⓑ -3
Ⓒ 25
Ⓓ -7
Ⓔ -100

Do NOT write your answers in this book. To open the answer sheet, scan the QR code or visit ***lumoslearning.com/a/6m017***

Chapter 2 → Lesson 11: Interpreting Absolute Value

1. What is the absolute value of the number represented by the dot plotted below?

Ⓐ -3
Ⓑ 1
Ⓒ 3
Ⓓ 4

2. What is the absolute value of the number represented by the dot plotted below?

Ⓐ -2
Ⓑ -1
Ⓒ 1
Ⓓ 2

3. Which symbol will make the following a true statement? $|-27|$ ________________ 19

Ⓐ >
Ⓑ <
Ⓒ =
Ⓓ ≤

4. Which symbol will make the following a true statement? -4 ________________ $-|-6|$

Ⓐ >
Ⓑ <
Ⓒ =
Ⓓ ≤

5. Which of the following numbers will make both inequalities true? –12 < x; |x| < 4

Ⓐ $x = -5$
Ⓑ $x = -1$
Ⓒ $x = -7$
Ⓓ $x = -4$

6. Ruth, who lives in Florida at an elevation of 30 meters, goes on a vacation to the Grand Cayman Islands, at an elevation of 24 meters, to go scuba diving at an elevation of –30 meters. Which elevation has the greatest absolute value?

Ⓐ 30 meters
Ⓑ 24 meters
Ⓒ –30 meters
Ⓓ Both 30 meters and –30 meters have the greater absolute value.

7. Connie, Julie, and Shelley's parents have encouraged them to save their money. Connie has an account balance of $215, Julie has –$100, and Shelley has –$250. Which inequality accurately represents the relative absolute values of each account.

Ⓐ |$215|<|-$100|<|-$250|
Ⓑ |-$250|<|-$100|<|$215|
Ⓒ |-$100|>|$215|>|-$250|
Ⓓ |-$100|<|$215|<|-$250|

8. The buoys on a certain lake mark the distance in meters from a center buoy. All buoys directly south are given negative numbers and all buoys directly north are given positive numbers. Betsy is located at buoy –6.2 and her brother Luis is located at buoy 6.5. Based on this information, which one of the following statements is not true.

Ⓐ Betsy is closer to the center buoy than Luis
Ⓑ Luis is 6.5 meters from the center buoy.
Ⓒ Luis is 0.2 meters farther from the center buoy than Betsy.
Ⓓ Betsy is 12.7 meters from Luis.

9. **Over the last three months Ophelia, Aaron, Nathan, and Rebecca recorded their weight. The table shows their initial and final weights.**

Name	Initial Weight, lbs	Final Weight, lbs
Ophelia	145	157
Aaron	178	163
Nathan	205	217
Rebecca	136	128

Whose weight had the least absolute change?

Ⓐ Ophelia
Ⓑ Aaron
Ⓒ Nathan
Ⓓ Rebecca

10. **The table below records the lowest and highest temperatures for four states.**

State	Lowest Temp, °F	Highest Temp, °F
New York	−52	108
Texas	−23	120
Florida	−2	109
North Carolina	−34	110

Which state has the broadest range of temperature extremes?

Ⓐ New York
Ⓑ Texas
Ⓒ Florida
Ⓓ North Carolina

11. **Sequence the numbers as they would fall on the number line in order from least to greatest. Enter the correct answers in the boxes given below.**

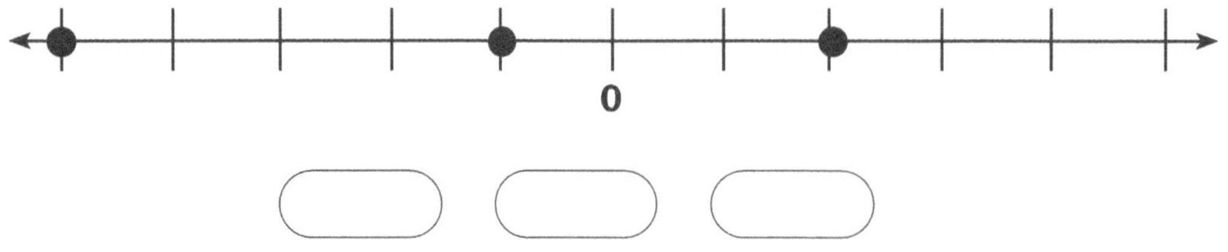

12. Which number highest absolute value in this number line?

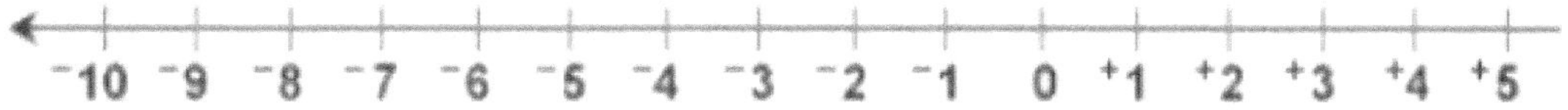

Ⓐ +4
Ⓑ -9
Ⓒ -5
Ⓓ +2
Ⓔ -10
Ⓕ -1

Do NOT write your answers in this book. To open the answer sheet, scan the QR code or visit ***lumoslearning.com/a/6m018***

Chapter 2 → Lesson 12: Comparisons of Absolute Value

1. **Which account balance represents the greatest debt?**
   \$20, –\$45, –\$5, \$10

   Ⓐ \$20
   Ⓑ –\$45
   Ⓒ –\$5
   Ⓓ \$10

2. **Which of the following is the warmest temperature?**
   5°F above zero, 6°F below zero, 10°F below zero, 2°F above zero

   Ⓐ 5°F above zero
   Ⓑ 6°F below zero
   Ⓒ 10°F below zero
   Ⓓ 2°F above zero

3. **Anneliese spent \$58.00 on music items and paid \$35.00 on a lay-away item. If she has –142.00 left in her account, how much did she start with?**

   Ⓐ \$–49.00
   Ⓑ \$93.00
   Ⓒ \$135.00
   Ⓓ –\$235.00

4. **The Murphys began their summer trip from home at an elevation of 439 feet. They drove to the mountains and climbed to an elevation of 1839 feet. After their visit on the top of the mountain, they drove down the mountainside 527 feet where they stopped for lunch. At the end of the day they spent the night at a campground at an elevation 264 feet higher than the restaurant. What is the difference in elevation between the campground and home?**

   Ⓐ 1048 feet
   Ⓑ 1137 feet
   Ⓒ 1576 feet
   Ⓓ 2012 feet

5. **Before Tonya went shopping she had \$165.00 in her account. She returned a jacket that cost \$46.50, bought 2 pairs of socks for \$5.99 each, and went to lunch and a movie for \$28.00. What is Tonya's account balance now?**

Ⓐ \$125.02
Ⓑ \$158.48
Ⓒ \$171.52
Ⓓ \$177.51

6. **The Casey quadruplets live in four different states. Dominik lives in Nantucket, Massachusetts at an elevation of 28 feet; Denzel lives in New Orleans, Louisiana at an elevation of 5.3 feet below sea level; Kaila lives in California near Death Valley at an elevation of 7 feet below sea level; and Malik lives in Nome, Alaska at an elevation of 20 feet. Who lives at the lowest elevation?**

Ⓐ Dominik
Ⓑ Denzel
Ⓒ Kaila
Ⓓ Malik

7. **One day in January the Casey quadruplets compared their location temperature.**

Location	Temperature, °C
Nome, Alaska	14.9 below zero
Nantucket, Massachusetts	5.1 below zero
New Orleans, Louisiana	6 above zero
Death Valley, California	15 above zero

Which location has the warmest temperature?

Ⓐ Nome, Alaska
Ⓑ Nantucket, Massachusetts
Ⓒ New Orleans, Louisiana
Ⓓ Death Valley, California

8. **Sato, her brother Ichiro, and two friends, Aran and Mio, went to a festival. Before they could board any ride they had to be taller than the wooden height checker at each ride. At one ride Sato was 4 inches taller, Ichiro was 2 inches shorter, Aran was 6 inches taller and Mio was 2.5 inches shorter than the wooden height checker. Who is the shortest person?**

Ⓐ Sato
Ⓑ Ichiro
Ⓒ Aran
Ⓓ Mio

9. Leary went to a sports store with $60.00 and bought a sweat shirt, athletic tape and baseball socks. If he received $3.72 in change, how much did his purchases cost?

Ⓐ $63.72
Ⓑ $61.18
Ⓒ $56.28
Ⓓ $47.28

10. Reilley owes $75.38 for his phone and $35.00 in dues. He expects to receive a credit of $26.16 for a wrong charge. If he currently has $629.00 in his account, what will be his account balance after all debts and credits are completed?

Ⓐ $713.22
Ⓑ $544.78
Ⓒ $518.62
Ⓓ $492.46

11. Select all numbers with an absolute value less than 20.

Ⓐ |-5|
Ⓑ |12|
Ⓒ |2|
Ⓓ |-18|
Ⓔ None of the above

12. Find the Value of |12| - |-11| = _______?

Do NOT write your answers in this book. To open the answer sheet, scan the QR code or visit ***lumoslearning.com/a/6m019***

Chapter 2 → Lesson 13: Coordinate Plane

1. **Ricky and Becca are going hiking. Below is the map that they are using. They start out at (−3.6, −2.6). They hike four units to the east and six units to the north. What are the coordinates of their new location? (Note: North is up on this map.)**

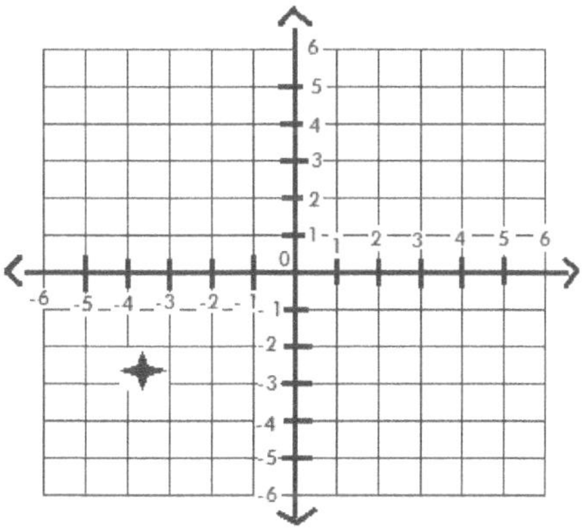

Ⓐ (0.6, 3.6)
Ⓑ (0.4, 3.4)
Ⓒ (0.4, 3.6)
Ⓓ (1.6, 3.6)

2. **The absolute value of the coordinates are (5, 8). What are the coordinates in Quadrant II?**

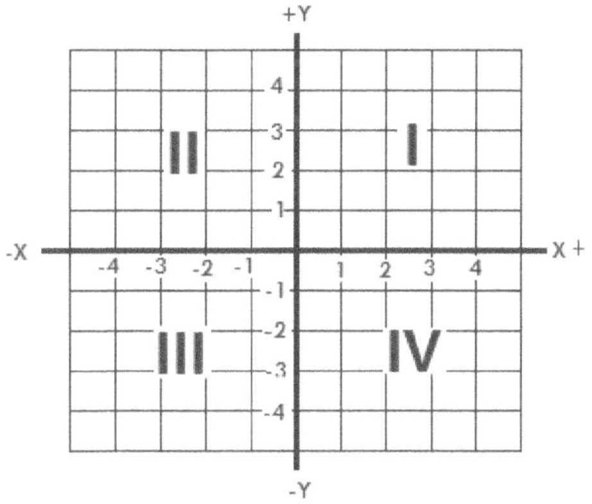

Ⓐ (−5, −8)
Ⓑ (5, −8)
Ⓒ (5, 8)
Ⓓ (−5, 8)

3. **What is the absolute value of the coordinates shown on the graph?**

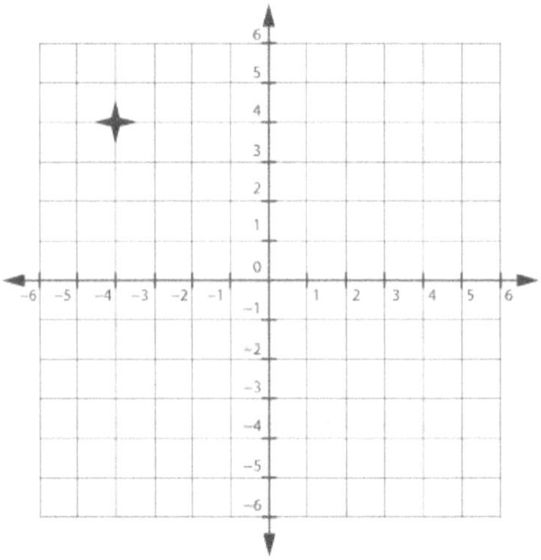

Ⓐ (−4, 4)
Ⓑ (4, 4)
Ⓒ (4, −4)
Ⓓ (−4, −4)

4. **Greg is in the jungle taking pictures of wildlife. He has to walk 5 units north and 3 units west to get back to his village. What are the coordinates of Greg's village?**
(Note: North is up on this map. Also, consider Greg's starting position as origin)

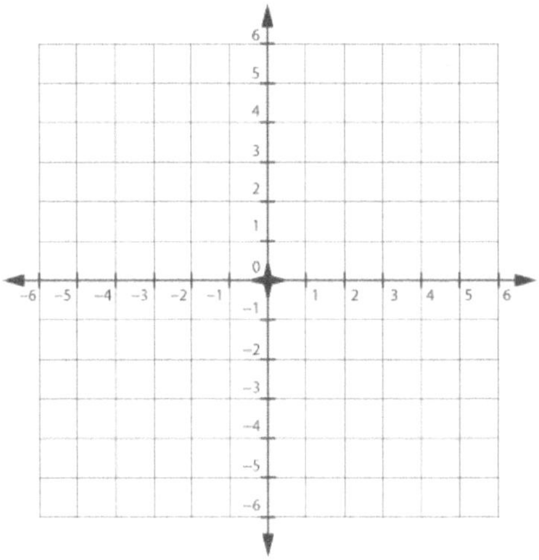

Ⓐ (−3, 5)
Ⓑ (3, 5)
Ⓒ (−3, −5)
Ⓓ (3, −5)

5. **How many units does Yolanda need to walk from point C to point D? (Note: North is up on this map.)**

Ⓐ 1 unit west and 7 units north
Ⓑ 1 unit east and 7 units south
Ⓒ 1 unit west and 7 units south
Ⓓ 7 units west and 1 unit south

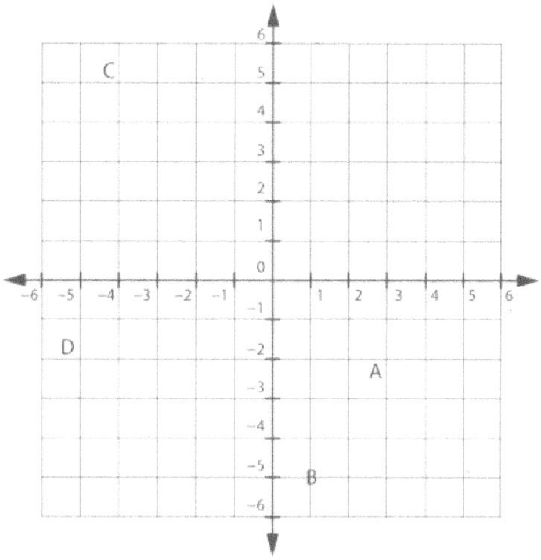

6. **What is the absolute value of Point D's coordinates?**

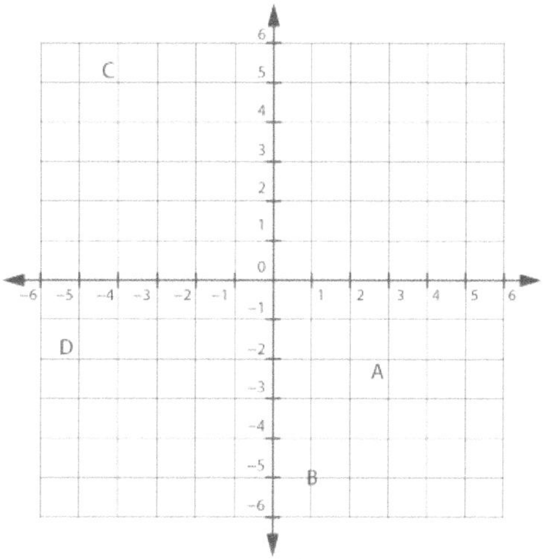

Ⓐ (−5, 2)
Ⓑ (5, −2)
Ⓒ (−5, −2)
Ⓓ (5, 2)

7. **There is a set of coordinates in Quadrant 4. The absolute value of x = 7. The absolute value of y = 3. What are the coordinates?**

Ⓐ (−7, −3)
Ⓑ (−3, −7)
Ⓒ (7, −3)
Ⓓ (−7, 3)

8. **There is a set of coordinates in Quadrant 2. The absolute value of x = 3. The absolute value of y = 9. What are the coordinates?**

Ⓐ (−9, −3)
Ⓑ (−3, −9)
Ⓒ (9, −3)
Ⓓ (−3, 9)

9. **The graph below shows the top of a mountain. If the value y = 0 is at sea level and the y-axis is measuring altitude, how far below sea level is the point (0, −6)?**

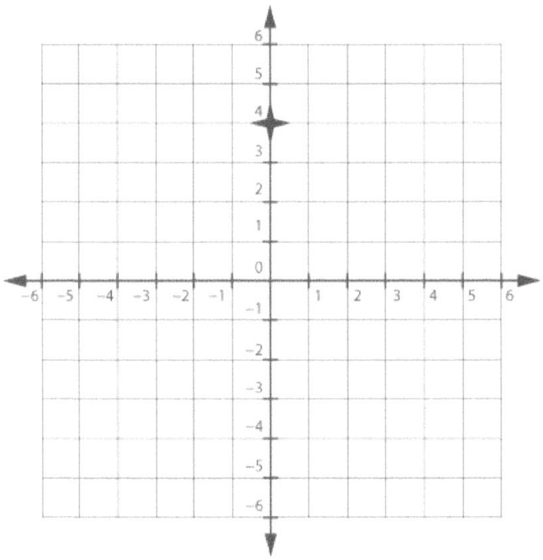

Ⓐ 10 units below sea level
Ⓑ 6 units above sea level
Ⓒ 6 units below sea level
Ⓓ 10 units above sea level

10. **The coordinates (−2, 8) are in Quadrant II. If the absolute values of x and y were to stay the same, but the point was moved to Quadrant IV, what would the coordinates be?**

Ⓐ (8, −2)
Ⓑ (−8, 2)
Ⓒ (−2, −8)
Ⓓ (2, −8)

11. **What is the distance between these points shown in the picture? Enter the answer in the box given below.**

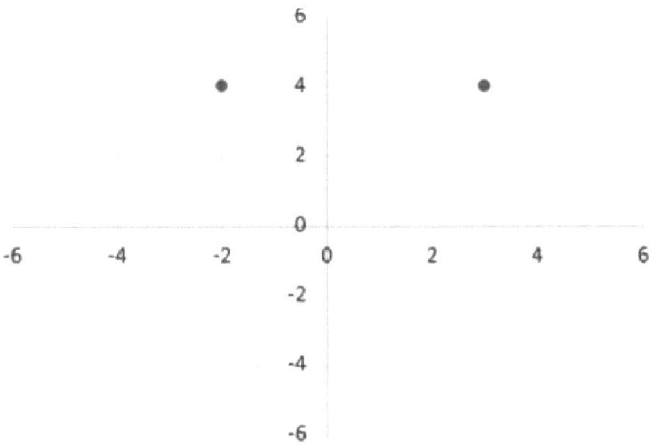

12. **Which of the following ordered pairs would fall in Quadrant I on the coordinate plane? Choose all that apply.**

Ⓐ (-4, 5)
Ⓑ (2, 3)
Ⓒ (2, -5)
Ⓓ (1, 1)
Ⓔ (3, 1)
Ⓕ (-3, -5)

13. **Circle the point that names the ordered pair (0, -8).**

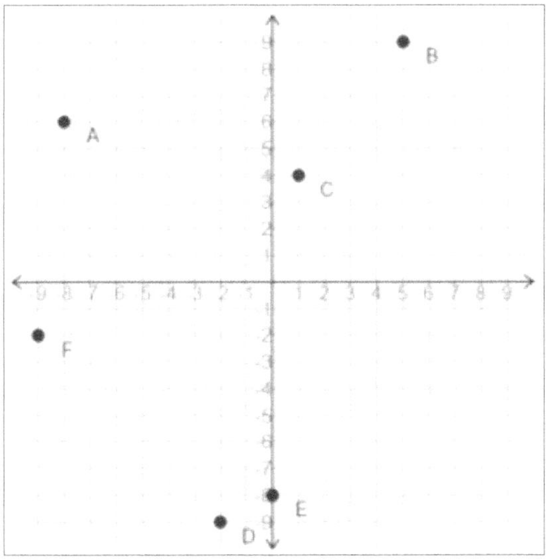

Point A	Point B	Point C
Point D	Point E	Point F

End of The Number System

Chapter 3

Expressions & Equations

Do NOT write your answers in this book. To open the answer sheet, scan the QR code or visit ***lumoslearning.com/a/6m020***

Lesson 1: Whole Number Exponents

1. **Evaluate: 5^3**

   Ⓐ 15
   Ⓑ 125
   Ⓒ 8
   Ⓓ 2

2. **Write the expression using an exponent: 2 * 2 * 2 * 2 * 2 * 2**

   Ⓐ 2 * 6
   Ⓑ 12
   Ⓒ 2^6
   Ⓓ 6^2

3. **Write the expression using an exponent: y * y * y * y**

   Ⓐ 4y
   Ⓑ y/4
   Ⓒ 4^y
   Ⓓ y^4

4. **Find the numerical value of the following expression: 11^1**

   Ⓐ 11
   Ⓑ 1
   Ⓒ 12
   Ⓓ 10

5. **Write an expression using exponents: 2 * 2 * m * m**

   Ⓐ 2(2m)
   Ⓑ 4m
   Ⓒ 2^2m^2
   Ⓓ 2/m

6. **Simplify: $4^3 * 4^2$**

Ⓐ 20
Ⓑ 9
Ⓒ 4^5
Ⓓ 20

7. **Simplify: $(b^2c)(bc^3)$**

Ⓐ 3b/4
Ⓑ 3b * 4
Ⓒ b^3c^4
Ⓓ bc

8. **Simplify: $(n^4x^2)^3$**

Ⓐ 12n*6x
Ⓑ $n^{12}x^6$
Ⓒ $n^{43}x^{23}$
Ⓓ n^7x^5

9. **Simplify: $7^4/7^2$**

Ⓐ 7^6
Ⓑ 7^3
Ⓒ 7^2
Ⓓ 7^4

10. **Simplify: $[(3^5)(3^2)]^4$**

Ⓐ 3^{28}
Ⓑ 3^{40}
Ⓒ 3^{10}
Ⓓ 3^{11}

11. **Select all numbers that would have a total value greater than 50.**

Ⓐ 6^2
Ⓑ 2^3
Ⓒ 5^2
Ⓓ 10^2
Ⓔ 4^4

12. **Find the numerical value of 8^4. Write your answer in standard form in the box.**

Do NOT write your answers in this book. To open the answer sheet, scan the QR code or visit ***lumoslearning.com/a/6m021***

Chapter 3 → Lesson 2: Expressions Involving Variables

1. **When the expression 3(n + 7) is evaluated for a given value of n, the result is 33. What is the value of n?**

   Ⓐ n = 4
   Ⓑ n = 5
   Ⓒ n = 21
   Ⓓ n = 120

2. **Which number is acting as a coefficient in this expression? 360 + 22x – 448**

   Ⓐ 360
   Ⓑ 22
   Ⓒ 448
   Ⓓ None of these

3. **Evaluate the following when n = 7: 5(n – 5)**

   Ⓐ 10
   Ⓑ −60
   Ⓒ 60
   Ⓓ 30

4. **For which of the following values of b does the expression 4b – 9 have a value between 90 and 100?**

   Ⓐ b = 104
   Ⓑ b = 26
   Ⓒ b = 48
   Ⓓ b = 24

5. **Evaluate the following when n = −4: [5n – 3n] + 2n**

   Ⓐ b = 16
   Ⓑ b = −20
   Ⓒ b = −16
   Ⓓ b = 0

6. Translate the following: "Four times a number n is equal to the difference between that number and 10"

Ⓐ 4n = 10 - n
Ⓑ 4 + n = 10*n
Ⓒ 4/n = n + 10
Ⓓ n - 4 = 10n

7. Evaluate 2y + 3y – y when y = 2.

Ⓐ 7
Ⓑ 8
Ⓒ 9
Ⓓ 10

8. Find the value of 2b – 4 + 6y when b = 2 and y = 3.

Ⓐ 16
Ⓑ 18
Ⓒ 0
Ⓓ −12

9. Translate the following, and then solve: "A number n times 16 is equal to 48."

Ⓐ 16n = 48, n = 4
Ⓑ 16n = 48, n = 8
Ⓒ 16 + n = 48, n = 32
Ⓓ 16n = 48, n = 3

10. For which value of x does 6x + 12 evaluate to 54?

Ⓐ x = 12
Ⓑ x = 9
Ⓒ x = 7
Ⓓ x = 6

11. Circle the answer choice that represents "12 less than f".

Ⓐ 12 - f
Ⓑ (12)f
Ⓒ f - 12

12. Which of the following represents the phrase "the quotient of 17 and q"? Select all the correct answers.

Ⓐ q ÷ 17
Ⓑ 17 ÷ q
Ⓒ 17/q
Ⓓ q/17

Do NOT write your answers in this book. To open the answer sheet, scan the QR code or visit ***lumoslearning.com/a/6m022***

Chapter 3 → Lesson 3: Identifying Expression Parts

1. Which of the following describes the expression 6(4–2) accurately?

Ⓐ Six and the difference of four and two.
Ⓑ The product of six and the sum of four and two.
Ⓒ The product of six and the difference of 4 and 2.
Ⓓ The quotient of six and the difference of four and two.

2. Which of the following describes the expression (8÷2)–10 accurately?

Ⓐ The quotient of eight and two subtracted from ten.
Ⓑ Ten less than the quotient of eight and two.
Ⓒ The difference of ten and the quotient of eight and two.
Ⓓ The product of eight and two minus ten.

3. What are the coefficients in the expression $(2x + 15)(9x - 3)$?

Ⓐ 2, 15, 9, −3
Ⓑ 15, 3
Ⓒ 15, −3
Ⓓ 2,9

4. What is the value of the greatest coefficient in the expression $4a^2 + 9a - 11b^2 + 15$?

Ⓐ 4
Ⓑ 9
Ⓒ 11
Ⓓ 15

5. How many factors are in the following expression, 4(6 + 8) × 3(2 – 5)?

Ⓐ 2
Ⓑ 4
Ⓒ 5
Ⓓ 6

6. **Which of the following describes the expression $4 \div (5 \times \frac{1}{2})$ accurately?**

Ⓐ The quotient of 4 and $5\frac{1}{2}$

Ⓑ 4 divided by the quotient of 5 and $\frac{1}{2}$

Ⓒ The quotient of 4 and the product of 5 and $\frac{1}{2}$

Ⓓ The product of 5 and $\frac{1}{2}$ divided by 4.

7. **Which of the following describes the expression (6+9) – 4 accurately?**

Ⓐ 4 subtracted from the sum of 6 and 9.
Ⓑ The sum of 6 and 9 subtracted from 4.
Ⓒ The difference of 6 and 9 less 4.
Ⓓ The sum of the quantity of 6 plus 9 and 4.

8. **Which term has the smallest coefficient in the expression $8x^4 + \frac{7}{8}x^3 - 2x^2 + x$?**

Ⓐ 1st term
Ⓑ 2nd term
Ⓒ 3rd term
Ⓓ 4th term

9. **Which of the following accurately represents "The product of 15 and the sum of 9 and 7."?**

Ⓐ 15 + (9 × 7)
Ⓑ (15 + 9) × 7
Ⓒ 15 ÷ (9 + 7)
Ⓓ 15(9 + 7)

10. **Which of the following describes the expression (12 ÷ 4) + [2 × (−2)] accurately?**

Ⓐ The sum of the quotient of 12 and 4 and the product of 2 and −2.
Ⓑ The quotient of 12 and 4 plus the difference of the product of 2 and 2.
Ⓒ The quotient of 4 and 12 plus the product of 2 and −2.
Ⓓ The difference of 2 times −2 and the quotient of 12 and 4.

11. **Which of the following means the same as 3(2+1)? Select all the correct answers.**

Ⓐ (2+1) + (2+1) + (2+1)
Ⓑ 6 + 3
Ⓒ 3 + 2 +1
Ⓓ (6 + 1) + (6 + 1) + (6 + 1)
Ⓔ None of the above

12. ______ is the same as (3+1) + (3+1). Circle the correct answer choice.

Ⓐ 2(3+1)
Ⓑ 3(2+1)
Ⓒ $(3+1)^2$

Do NOT write your answers in this book. To open the answer sheet, scan the QR code or visit ***lumoslearning.com/a/6m023***

Chapter 3 → Lesson 4: Evaluating Expressions

1. What is the value of y in the equation $y = 3x - 13$, when x = 6?

Ⓐ 5
Ⓑ −4
Ⓒ −1
Ⓓ 4

2. What is the value of y in the equation $y = \frac{1}{4}x \div 2$, when x = 32?

Ⓐ 2
Ⓑ 4
Ⓒ 8
Ⓓ 10

3. Evaluate the following expression when a = 3 and b = −8: $3a^2 - 7b$

Ⓐ −44
Ⓑ −29
Ⓒ 68
Ⓓ 83

4. Evaluate the following expression when v = −2 and w = 155: $6v^3 + \frac{4}{5}w$

Ⓐ 17
Ⓑ 68
Ⓒ 76
Ⓓ 172

5. Evaluate the following expression when c = −3 and d = 2: $\frac{6}{d} - 10c - c^4$

Ⓐ −108
Ⓑ −48
Ⓒ 39
Ⓓ 114

6. Use the formula $V = s^3$ to find the volume of a cube with a side length of 2 cm.

Ⓐ 4 cm^2
Ⓑ 6 cm^3
Ⓒ 8 cm^3
Ⓓ 9 cm^3

7. Use the formula $A = l \times w$ to find the area of a rectangle with a length of 4.5 feet and a width of 7.3 feet.

Ⓐ 28.15 ft^2
Ⓑ 30.35 ft^2
Ⓒ 32.35 ft^2
Ⓓ 32.85 ft^2

8. Leah found a cylindrical container with a radius of $1\frac{1}{2}$ inches and a height of 11 inches. Use the formula $V = \pi r^2 h$, where $\pi = 3.14$, r is the radius and h is the height, to find the volume of the container. Round to the nearest hundredth.

Ⓐ 51.81 in^3
Ⓑ 77.72 in^3
Ⓒ 103.62 in^3
Ⓓ 566.28 in^3

9. What is the area of a square with a side of $\frac{3}{4}$ meter, $A = s^2$, where A is the area of the square, s it's length.

Ⓐ $\frac{3}{16}$ m^2

Ⓑ $\frac{9}{16}$ m^2

Ⓒ $\frac{3}{4}$ m^2

Ⓓ $\frac{9}{4}$ m^2

10. The area of the base of a square pyramid is 10 ft^2 and the height is 6 ft. What is the volume of the pyramid using the formula $V = \frac{1}{3}\beta h$, where β is the area of the base and h is the height?

Ⓐ 20 ft^3
Ⓑ 60 ft^3
Ⓒ 200 ft^3
Ⓓ 600 ft^3

11. Circle the answer that fits the 'x' in this equation $\frac{1200}{x} = 30$.

Ⓐ 4
Ⓑ 40
Ⓒ 400

12. Select the equations in which j = 7. Choose all that apply.

Ⓐ $3j - 4 = 25$
Ⓑ $56 - j = 49$
Ⓒ $3j + 4 = 25$
Ⓓ $j^3 = 343$

Do NOT write your answers in this book. To open the answer sheet, scan the QR code or visit ***lumoslearning.com/a/6m024***

Chapter 3 → Lesson 5: Writing Equivalent Expressions

1. What is an equivalent expression for 3n – 12?

Ⓐ 3n + 1
Ⓑ 3n + 4
Ⓒ 3n – 4
Ⓓ 3(n – 4)

2. Simplify 2n – 7n to create an equivalent expression.

Ⓐ 5n
Ⓑ –5n
Ⓒ –n(2 – 7)
Ⓓ n(5)

3. Which expression is equivalent to 5y + 2z – 3y + z?

Ⓐ z
Ⓑ 2y + 3z
Ⓒ yz
Ⓓ 11yz

4. Which expression is equivalent to 8a + 9 – 3(a + 4)?

Ⓐ 32 – 3a
Ⓑ a
Ⓒ 24a
Ⓓ 5a – 3

5. Which inequality has the same solution set as 3(q + 6) > 11?

Ⓐ 3q + 18 < 11
Ⓑ 3q + 18 > 11
Ⓒ 3q + 6 > 11
Ⓓ q + 18 > 11

6. Which inequality has the same solution set as $10 < q + q + q + q + q - 5$?

Ⓐ $10 > 5q + 5$
Ⓑ $10 < 5q + 5$
Ⓒ $10 > 5q - 5$
Ⓓ $10 < 5q - 5$

7. Simplify the following equation:
$u + u + u + u - p + p + p - r = 55$

Ⓐ $4u + 2p - r = 55$
Ⓑ $4u + p - r = 55$
Ⓒ $4u - 3p + r = 55$
Ⓓ $4u - 3p - r = 55$

8. $36x - 12 = 108$ has the same solution(s) as ______________ .

Ⓐ $3(3x - 12) = 108$
Ⓑ $12(3x - 1) = 108$
Ⓒ $3(12x - 12) = 108$
Ⓓ $12(x - 1) = 108$

9. Why is the expression $5(3x + 2)$ equivalent to $15x + 10$?

Ⓐ The 5 has been divided into each term in parentheses.
Ⓑ The 5 was distributed using the Distributive Property.
Ⓒ The 5 was distributed using the Associative Property.
Ⓓ The expressions are not equal.

10. Which expression is equivalent to $5b - 9c - 2(4b + c)$?

Ⓐ $-3b + 7c$
Ⓑ $-3b - 7c$
Ⓒ $-3b - 11c$
Ⓓ $-3b + 11c$

11. Write the correct equation for the following expression. "3 less than the product of 4 and 5."

12. Circle the answer that represents 2(4+3b).

Ⓐ 8 + 6b
Ⓑ 6 + 8b
Ⓒ 6b - 8

Do NOT write your answers in this book. To open the answer sheet, scan the QR code or visit ***lumoslearning.com/a/6m025***

Chapter 3 → Lesson 6: Identifying Equivalent Expressions

1. **Which two expressions are equivalent?**

   Ⓐ $(\frac{5}{25})x$ and $(\frac{1}{3})x$
   Ⓑ $(\frac{5}{25})x$ and $(\frac{1}{5})x$
   Ⓒ $(\frac{5}{25})x$ and $(\frac{1}{4})x$
   Ⓓ $(\frac{5}{25})x$ and $(\frac{1}{6})x$

2. **Which two expressions are equivalent?**

   Ⓐ 7 + 21v and 2(5 + 3v)
   Ⓑ 7 + 21v and 3(4 + 7v)
   Ⓒ 7 + 21v and 7(1 + 3v)
   Ⓓ 7 + 21v and 7(7 +21v)

3. **Which two expressions are equivalent?**

   Ⓐ $\frac{32p}{2}$ and 17p
   Ⓑ $\frac{32p}{2}$ and 18p
   Ⓒ $\frac{32p}{2}$ and 16p
   Ⓓ $\frac{32p}{2}$ and 14p

4. **Which two expressions are equivalent?**

   Ⓐ 17(3m + 4) and 51m + 68
   Ⓑ 17(3m + 4) and 51m + 67
   Ⓒ 17(3m + 4) and 51m − 68
   Ⓓ 17(3m + 4) and 47m + 51

5. **Which two expressions are equivalent?**

Ⓐ $\frac{64k}{4}$ and 4k
Ⓑ $\frac{64k}{4}$ and 14k
Ⓒ $\frac{64k}{4}$ and 16k
Ⓓ $\frac{64k}{4}$ and 15k

6. **575d – 100 is equivalent to:**

Ⓐ 25(23d – 4)
Ⓑ 25(22d – 4)
Ⓒ 25(23d + 4)
Ⓓ 25(25d – 4)

7. **(800 + 444y)/4 is equivalent to:**

Ⓐ 200 + 44y
Ⓑ 800 + 111y
Ⓒ 200 + 111y
Ⓓ 200 – 111y

8. **5(19 –8y) is equivalent to:**

Ⓐ 95 – 35y
Ⓑ 95 + 40y
Ⓒ 85 – 40y
Ⓓ 95 – 40y

9. **The expression 3(26p – 7 + 14h) is equivalent to:**

Ⓐ 78 – 21 + 42
Ⓑ 78p + 21 + 42h
Ⓒ 78p – 21 + 42
Ⓓ 78p – 21 + 42h

10. **5(6x + 17y – 9z) is equivalent to:**

Ⓐ 30x + 82y – 45z
Ⓑ 20x + 85y – 40z
Ⓒ 30x + 85y – 45z
Ⓓ 30x – 85y + 45z

11. Which of the following equations represents 4(2 + 1c)? Choose all that apply.

Ⓐ 6 + 4c
Ⓑ 6 + 5c
Ⓒ 8 x 4c
Ⓓ 8 + 4c
Ⓔ (4 x 2) + (4 x 1c)
Ⓕ (4 x 2) x (4 x 1c)

12. [(8y) + (8y) + (8y)] ÷ 2 = _____ . Simplify the expression and write the answer in the box.

Do NOT write your answers in this book. To open the answer sheet, scan the QR code or visit ***lumoslearning.com/a/6m026***

Chapter 3 → Lesson 7: Equations and Inequalities

1. **How many positive whole number solutions (values for x) does this inequality have?**
   $x \le 20$

   Ⓐ 19
   Ⓑ 20
   Ⓒ 21
   Ⓓ Infinite

2. **Which of the following correctly shows the number sentence that the following words describe? *17 is less than or equal to the product of 6 and q.***

   Ⓐ $17 \le 6q$
   Ⓑ $17 \le 6 - q$
   Ⓒ $17 < 6q$
   Ⓓ $17 \ge 6q$

3. **Which of the following correctly shows the number sentence that the following words describe? *The quotient of d and 5 is 15.***

   Ⓐ $\frac{5}{d} = 15$
   Ⓑ $5d = 15$
   Ⓒ $\frac{d}{5} = 15$
   Ⓓ $d - 5 = 15$

4. **Which of the following correctly shows the number sentence that the following words describe? *Three times the quantity u – 4 is less than 17***

   Ⓐ $3(u - 4) > 17$
   Ⓑ $3(u - 4) < 17$
   Ⓒ $3(u - 4) \le 17$
   Ⓓ $3(u - 4) \ge 17$

5. **Which of the following correctly shows the number sentence that the following words describe? *The difference between z and the quantity 7 minus r is 54.***

Ⓐ $z - 7 - r = 54$
Ⓑ $z + 7 - r = 54$
Ⓒ $z + (7 - r) = 54$
Ⓓ $z - (7 - r) = 54$

6. **Which of the following correctly shows the number sentence that the following words describe? The square of the sum of 6 and b is greater than 10.**

Ⓐ $(6 + b)^2 > 10$
Ⓑ $6^2 + b^2 > 10$
Ⓒ $(6 + b)^2 = 10$
Ⓓ $(6 + b)^2 < 10$

7. **Which of the following correctly shows the number sentence that the following words describe? *16 less than the product of 5 and h is 21.***

Ⓐ $16 - 5h = 21$
Ⓑ $5h - 16 = 21$
Ⓒ $16 - (5 + h) = 21$
Ⓓ $16 < 5h + 21$

8. **Which of the following correctly shows the number sentence that the following words describe? *8 times the quantity 2x – 7 is greater than 5 times the quantity 3x + 9.***

Ⓐ $8(2x) - 7 > 5(3x) + 9$
Ⓑ $8(2x - 7) \geq 5(3x + 9)$
Ⓒ $8(2x - 7) > 5(3x + 9)$
Ⓓ $8(2x - 7) < 5(3x + 9)$

9. **A batting cage offers 8 pitches for a quarter. Raul has $1.50. Which expression could be used to calculate how many pitches Raul could get for his money?**

Ⓐ $1.50 x 8
Ⓑ $1.50 ÷ 8
Ⓒ ($1.50 ÷ $0.25) x 8
Ⓓ ($1.50 ÷ $0.25)

10. For which of the following values of x is this inequality true?
500 – 3x > 80

Ⓐ x = 140
Ⓑ x = 150
Ⓒ x = 210
Ⓓ x = 120

11. Circle the answer that correctly represents "d" in this equation.

3d + 4 > 17

Ⓐ 5
Ⓑ 2
Ⓒ 4

12. Select all values that could correctly represent b in the equation.

3b + 2 < 15

Ⓐ 1
Ⓑ 2
Ⓒ 4
Ⓓ 5
Ⓔ 7

Do NOT write your answers in this book. To open the answer sheet, scan the QR code or visit ***lumoslearning.com/a/6m027***

Chapter 3 → Lesson 8: Modeling with Expressions

1. **Roula had 117 gumballs. Amy had x less than $\frac{1}{2}$ the amount that Roula had. Which expression shows how many gumballs Amy had?**

   Ⓐ $117 - \frac{x}{2}$
   Ⓑ $(\frac{1}{2})(117) - x$.
   Ⓒ $117 - 2x$
   Ⓓ $2x + 117$

2. **Benny earned $20.00 for weeding the garden. He also earned c dollars for mowing the lawn. Then he spent x dollars at the candy store. Which expression best represents this situation?**

   Ⓐ $20 − c − x
   Ⓑ $20 + c − x
   Ⓒ $20 + c + x
   Ⓓ $20 + $\frac{x}{c}$

3. **Clinton loves to cook. He makes a total of 23 different items. Clinton makes 6 different desserts, 12 appetizers, and x main courses. Which equation represents the total amount of food that Clinton cooked?**

   Ⓐ $6 - 12 + x = 23$
   Ⓑ $6 + 12 + x = 23$
   Ⓒ $18 - x = 23$
   Ⓓ $6 + 12 - x = 23$

4. **Simon has read 694 pages over the summer by reading 3 different books. He read 129 pages in the first book and he read 284 pages in the second book. Which equation shows how to figure out how many pages he read in the third book?**

   Ⓐ $694 = 129 + 284 - y$
   Ⓑ $694= 413 - y$
   Ⓒ $694 = 129 - 284 + y$
   Ⓓ $694 = 129 + 284 + y$

5. **Jimmy had $45.00. He spent all of the money on a hat and a pair of jeans. He spent $19.00 on the pair of jeans and x dollars on the hat. Which of the following equations is true?**

Ⓐ $45.00 + x = $19.00
Ⓑ $45.00 − x = $19.00
Ⓒ x − $19.00 = $45.00
Ⓓ $19.00 + 45.00 = x

6. **Janie had 54 stamps. She gave away t stamps. She then got back twice as many as she had given away. Which expression shows how many stamps Janie has now?**

Ⓐ 54 − t
Ⓑ 54 + t
Ⓒ 54 − 2t
Ⓓ 2t − 54

7. **The library has 2,500 books. The librarian wants to purchase x more books for the library. The director decides to buy twice as many as the librarian requested. How many books will the library have if the director purchases the number of books he wants?**

Ⓐ 2,500 + x
Ⓑ 2,500 + 2x
Ⓒ 2,500 − x
Ⓓ 2,500 − 2x

8. **There are 24 boys and 29 girls (not including Claire) attending Claire's birthday party. Which equation shows how many cupcakes Claire needs to have so that everyone, including herself, will have a cupcake?**

Ⓐ 24 − 29 = c
Ⓑ 24 + 29 = c
Ⓒ 24 + 30 = c
Ⓓ 24 + c = 29

9. **Heidi collects dolls. She had 172 dolls in her collection. Heidi acquired x more dolls from a friend. She then bought twice as many dolls as she acquired from her friend from a yard sale. She now has 184 dolls in her collection. Which equation is true?**

Ⓐ 172 + 3x = 184
Ⓑ 172 − x + 2x = 184
Ⓒ 172(3x) = 184
Ⓓ 172 − 3x = 184

10. Crystal grew 16 tomato plants. Each plant grew 10 tomatoes. She sold x of the tomatoes she had grown. Crystal has 54 tomatoes left for herself. Which equation is true?

Ⓐ $10(16 + x) = 54$
Ⓑ $16(10 - x) = 54$
Ⓒ $16 + x = 540$
Ⓓ $160 - x = 54$

11. Circle the answer that represents the statement 43 is greater than q correctly.

Ⓐ $q > 43$
Ⓑ $43 + q$
Ⓒ $43 > q$

12. Choose the box(es) that demonstrate that x is a number greater than 5. Choose all that apply.

Ⓐ $x < 5$
Ⓑ $x > 5$
Ⓒ $x + 5$
Ⓓ $x - 5$
Ⓔ $5 < x$

Do NOT write your answers in this book. To open the answer sheet, scan the QR code or visit ***lumoslearning.com/a/6m028***

Chapter 3 → Lesson 9: Solving One-Step Equations

1. Which of the following equations describes this function?

X	Y
13	104
17	136
20	160
9	72

Ⓐ y = 18x
Ⓑ y = x + 4
Ⓒ y = x + 32
Ⓓ y = 8x

2. What is the value of x?

– 7x = 56

Ⓐ x = −7
Ⓑ x = 8
Ⓒ x = −49
Ⓓ x = −8

3. Does this table show a linear relationship between x and y?

X	Y
13	169
15	225
12	144
20	400
16	256
7	49
8	64

Ⓐ yes
Ⓑ no
Ⓒ yes, but only when x is positive
Ⓓ yes, but only when y is a perfect square

4. **Find the value of z:** $\frac{z}{5} = 20$

Ⓐ 100
Ⓑ 4
Ⓒ 15
Ⓓ 25

5. **Find the value of y:** $\frac{y}{3} = 12$

Ⓐ 9
Ⓑ 15
Ⓒ 36
Ⓓ 4

6. **Find the value of p in the following expression.**
13 + p = 39

Ⓐ 3
Ⓑ 26
Ⓒ 507
Ⓓ 52

7. **Find the value of *w* in the following expression.**
6*w* = 54

Ⓐ 48
Ⓑ 60
Ⓒ 324
Ⓓ 9

8. **Find the value of *h* in the following expression.**
***h* – 4 = 20**

Ⓐ 24
Ⓑ 16
Ⓒ –80
Ⓓ –5

9. **Find the value of p in the following expression.**
72 + p = 108

Ⓐ p = 30
Ⓑ p = 36
Ⓒ p = 180
Ⓓ p = 40

10. **Find the value of n in the following expression.**
428 – n = 120

Ⓐ n = −548
Ⓑ n = 308
Ⓒ n = −308
Ⓓ n = 548

11. **Circle the answer that represents m correctly in the equation 4m = 16.**

Ⓐ m = 4
Ⓑ m = 3
Ⓒ m = 12

12. **Select the equations in which k = 8. Select all the correct answers.**

Ⓐ 3(6 + k) = 42
Ⓑ ((4) 7) / k =14
Ⓒ 8k - 4 = 60
Ⓓ 7(k / 2) =35

Do NOT write your answers in this book. To open the answer sheet, scan the QR code or visit ***lumoslearning.com/a/6m029***

Chapter 3 → Lesson 10: Representing Inequalities

1. A second grade class raised caterpillars. They had 12 caterpillars. Less than half of the caterpillars turned into butterflies. Which inequality shows how many caterpillars turned into butterflies?

Ⓐ $x < 6$
Ⓑ $x > 6$
Ⓒ $x \leq 6$
Ⓓ $x \geq 6$

2. Elliot has at least 5 favorite foods. How many favorite foods could Elliot have?

Ⓐ 4
Ⓑ 2
Ⓒ none
Ⓓ an infinite number

3. Julie has a box full of crayons. Her box of crayons has 549 crayons and at least 8 of them are red. Which inequality represents how many crayons could be red?

Ⓐ $x \geq 549$
Ⓑ $8 \geq x \geq 549$
Ⓒ $8 \geq x$
Ⓓ $8 \leq x \leq 549$

4. Five times a number is greater than that number minus 17 is represented as ____________ .

Ⓐ $5x > x - 17$
Ⓑ $x + 5 > x - 17$
Ⓒ $5x < x - 17$
Ⓓ $5x > x + 17$

5. "A number divided by five minus five is less than negative four" is represented as__________ .

Ⓐ $5x - 5 < -4$
Ⓑ $x/5 - 5 < -4$
Ⓒ $x/5 - 5 > -4$
Ⓓ $x/5 - 5 < 4$

6. **"Three times the sum of six times a number and three is less than 27." is represented as ________________.**

Ⓐ $6x + 3 < 27$
Ⓑ $3(6)x + 3 > 27$
Ⓒ $3(6x + 3) < 27$
Ⓓ $6x + 3(3) < 27$

7. **How would $x > 3$ be represented on a number line?**

Ⓐ The number line would show an open circle over three with an arrow pointing to the left.
Ⓑ The number line would show an open circle over three with an arrow pointing to the right.
Ⓒ The number line would show a closed circle over three with an arrow pointing to the right.
Ⓓ The number line would show a closed circle over three with an arrow pointing to the left.

8. **Amy and Joey each have jellybeans. The amount Amy has is 3 times the amount that Joey has. There are at least 44 jellybeans between them. Which inequality would help you figure out how many jellybeans Amy and Joey each have?**

Ⓐ $x + 3x \geq 44$
Ⓑ $3x \geq 44$
Ⓒ $3 + x \geq 44$
Ⓓ $x + 3x \leq 44$

9. **Sandra is a lawyer. She is working on x number of cases. She gets 8 more cases to work on. She now has more than 29 cases that she is working on. Which inequality could be used to figure out how many cases Sandra is working on?**

Ⓐ $8x > 29$
Ⓑ $x + 8 < 29$
Ⓒ $x + 8 > 29$
Ⓓ $x - 8 < 29$

10. **There are 25 beehives on a farm. There are the same number of bees in each hive. The total number of bees on the farm is greater than 800. Which inequality could be used to figure out how many bees are in each hive?**

Ⓐ $25/x > 800$
Ⓑ $25x < 800$
Ⓒ $25x > 800$
Ⓓ $25 + x > 800$

11. Circle the answer that represents the inequality shown on the number line below.

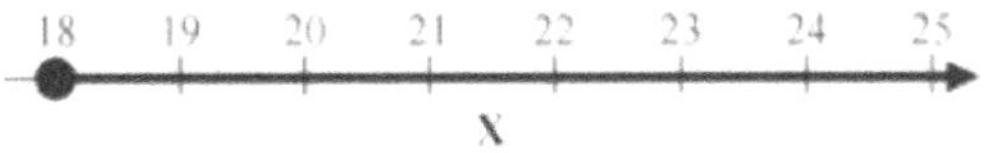

Ⓐ $x \geq 18$
Ⓑ $x < 18$
Ⓒ $x = 18$

12. Choose the box that best represents the inequality on the line below.

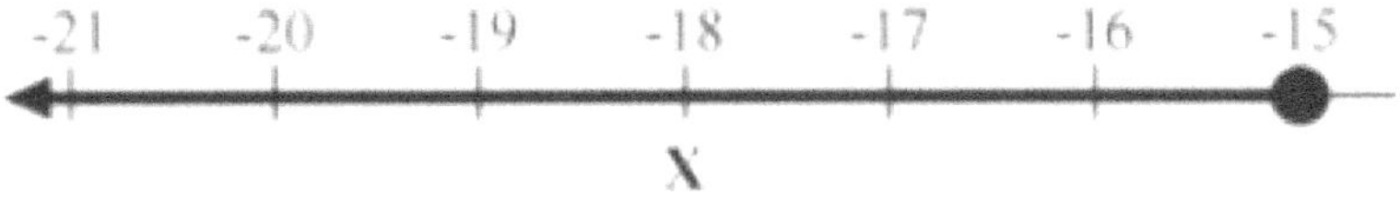

Ⓐ $x \geq 15$
Ⓑ $x \geq -15$
Ⓒ $x > -15$
Ⓓ $x < -15$
Ⓔ $x \leq -15$

13. A movie is rated PG13 meaning that one must be at least 13-years old to watch the movie. The sign in the lobby of the theater reads

PG13 Viewers
must be ≤ 13

Nami thinks the sign is wrong, but her friend Tai disagrees and finds nothing wrong with the sign. Who is correct and why? Enter your answer in the box.

14. "A number divided by 5 is at most -23" is represented as ________________.
Write the inequality and explain.

Do NOT write your answers in this book. To open the answer sheet, scan the QR code or visit ***lumoslearning.com/a/6m030***

Chapter 3 → Lesson 11: Quantitative Relationships

1. **Logan loves candy! He goes to the store and sees that the bulk candy is \$0.79 a pound. Logan wants to buy p pounds of candy and needs to know how much money (m) he needs. Which equation would be used to figure out how much money Logan needs?**

   Ⓐ m = 0.79 ÷ p
   Ⓑ m = 0.79(p)
   Ⓒ 0.79 = m(p)
   Ⓓ m = 0.79 + p

2. **Logan loves candy! He goes to the store and sees that the bulk candy is \$0.84 a pound. Logan wants to buy 3 pounds of candy. Using the equation m = 0.84(p), figure out how much money (m) Logan needs.**

   Ⓐ \$1.68
   Ⓑ \$2.52
   Ⓒ \$2.25
   Ⓓ \$2.54

3. **Norman is going on a road trip. He has to purchase gas so that he can make it to his first destination. Gas is \$3.55 a gallon. Norman gets g gallons. Which equation would Norman use to figure out how much money (t) it cost to get the gas?**

   Ⓐ t = g(3.55)
   Ⓑ t = g ÷ 3.55
   Ⓒ t = 3.55 ÷ g
   Ⓓ t = g + 3.55

4. **Norman is going on a road trip. He has to purchase gas so that he can make it to his first destination. Gas is \$3.58 a gallon. Norman needs to get 13 gallons. Using the expression t = g(3.58), figure out how much Norman will spend on gas.**

   Ⓐ \$46.15
   Ⓑ \$39.54
   Ⓒ \$46.45
   Ⓓ \$46.54

5. Penny planned a picnic for her whole family. It has been very hot outside, so she needs a lot of lemonade to make sure no one is thirsty. There are 60 ounces in each bottle. Penny purchased b bottles of lemonade. She wants to figure out the total number of ounces (o) of lemonade she has. Which equation should she use?

Ⓐ b = 60(o)
Ⓑ 60 = o × b
Ⓒ 60(b) = o
Ⓓ 60 = b ÷ o

6. Ethan is playing basketball in a tournament. Each game lasts 24 minutes. Ethan has 5 games to play. Which general equation could he use to help him figure out the total number of minutes that he played? Let t = the total time, g = the number of games, and m = the time per game.

Ⓐ t = g + m
Ⓑ t = g(m)
Ⓒ t = g − m
Ⓓ t = g ÷ m

7. The Spencers built a new house. They want to plant trees around their house. They want to plant 8 trees in the front yard and 17 in the backyard. The trees that the Spencer's want to plant cost $46 each. Could they use the equation t = c(n) where t is the total cost, c is the cost per tree, and n is the number of trees purchased, to figure out the cost to purchase trees for both the front and the back yards?

Ⓐ No, because the variables represent only two specific numbers that will never change.
Ⓑ Yes, because the variables represent only two specific numbers that will never change.
Ⓒ No, because the variables can be filled in with any number.
Ⓓ Yes, because the variables can be filled in with any number.

8. Liz is a florist. She is putting together b bouquets for a party. Each bouquet is going to have f sunflowers in it. The sunflowers cost $3 each. Which equation can Liz use to figure out the total cost (c) of the sunflowers in the bouquets?

Ⓐ c = 3(bf)
Ⓑ c = bf ÷ 3
Ⓒ c = 3b + f
Ⓓ c = 3(b + f)

9. **Liz is a florist. She is putting together 5 bouquets for a party. Each bouquet is going to have 6 sunflowers in it. The sunflowers cost $3 each. Using the equation c = 3(bf), figure out how much Liz will charge for the bouquets.**

Ⓐ $90
Ⓑ $30
Ⓒ $18
Ⓓ $80

10. **Hen B lays 4 times as many eggs as Hen A. (Let a = Number of eggs Hen A lays, b = Number of eggs Hen B lays)**

Hen A	Hen B
2	8
4	16
7	28
11	44

Which equation represents this scenario?

Ⓐ b = 4 ÷ a
Ⓑ b = 4 + a
Ⓒ b = 4a
Ⓓ b = 4a − 4

11. **Which of the following represent balloons that have a cost of $2.50 for a quantity of 10? Choose all that apply.**

Ⓐ 20 balloons cost $1.25
Ⓑ 100 balloons $25.00
Ⓒ 30 balloons cost $7.50
Ⓓ 80 balloons cost $8.00

12. **Sodas cost $1.25 at the vending machine. Complete the table to show the quantity and total cost of sodas purchased.**

Day	Money Spent on Sodas	Sodas Purchased	Price per Soda
Monday		24	$1.25
Wednesday	$57.50		$1.25
Friday	$41.25		$1.25

End of Expressions & Equations

Chapter 4

Geometry

Do NOT write your answers in this book. To open the answer sheet, scan the QR code or visit ***lumoslearning.com/a/6m031***

Lesson 1: Area

1. **What is the area of the figure below?**

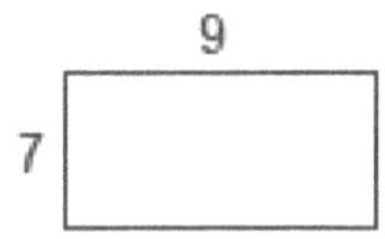

Ⓐ 16 square units
Ⓑ 63 square units
Ⓒ 32 square units
Ⓓ 45 square units

2. **What is the area of the figure below?**

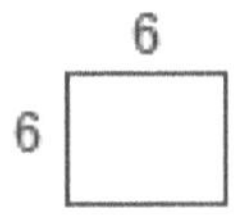

Ⓐ 12 square units
Ⓑ 24 square units
Ⓒ 18 square units
Ⓓ 36 square units

3. **What is the area of the figure below? (Assume that the vertical height of the parallelogram is 3 units.)**

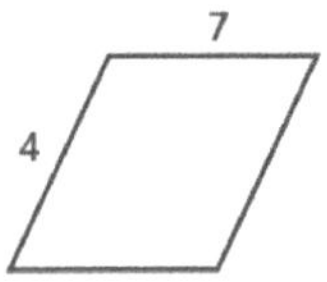

Ⓐ 28 square units
Ⓑ 12 square units
Ⓒ 14 square units
Ⓓ 21 square units

4. **The figure shows a small square inside a larger square. What is the area of the shaded portion of the figure below?**

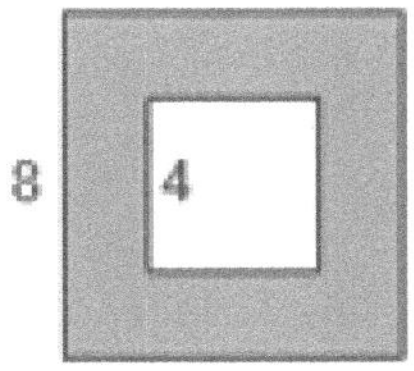

Ⓐ 64 square units
Ⓑ 48 square units
Ⓒ 16 square units
Ⓓ 80 square units

5. **What is the area of the figure below?**

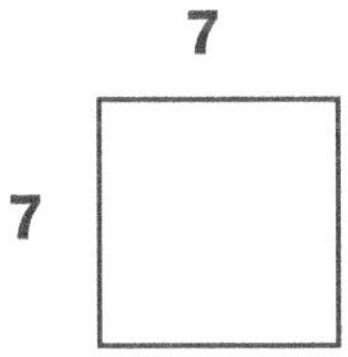

Ⓐ 14 square units
Ⓑ 28 square units
Ⓒ 49 square units
Ⓓ 21 square units

6. **What is the area of the figure below? (Assume that the vertical height of the triangle is 2.8 units)**

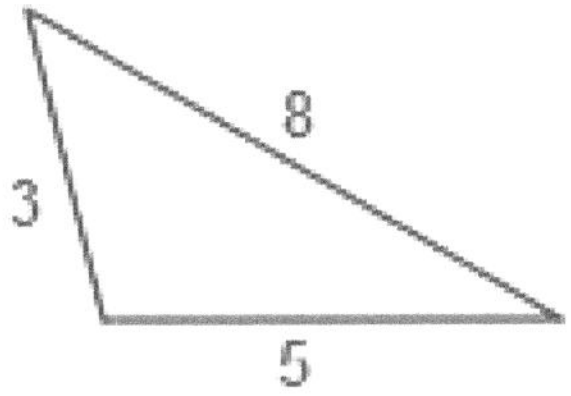

Ⓐ 15 square units
Ⓑ 7 square units
Ⓒ 14 square units
Ⓓ 40 square units

7. **What is the area of the figure below?**

5

12

Ⓐ 60 square units
Ⓑ 17 square units
Ⓒ 34 square units
Ⓓ 7 square units

8. **What is the area of the figure shown below? The vertical height is 6 units.**

19

7

Ⓐ 133 square units
Ⓑ 26 square units
Ⓒ 52 square units
Ⓓ 114 square units

9. **What is the area of the gray part of the squares below?.**

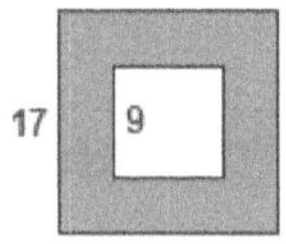

Ⓐ 289 square units
Ⓑ 81 square units
Ⓒ 208 square units
Ⓓ 370 square units

10. **What is the area of a triangle with a base of 20 feet and a vertical height of 40 feet?**

Ⓐ A = 200 square ft.
Ⓑ A = 800 square ft.
Ⓒ A = 400 square ft.
Ⓓ A = 600 square ft.

11. Calculate the area of the triangle shown below and write the answer in the box.

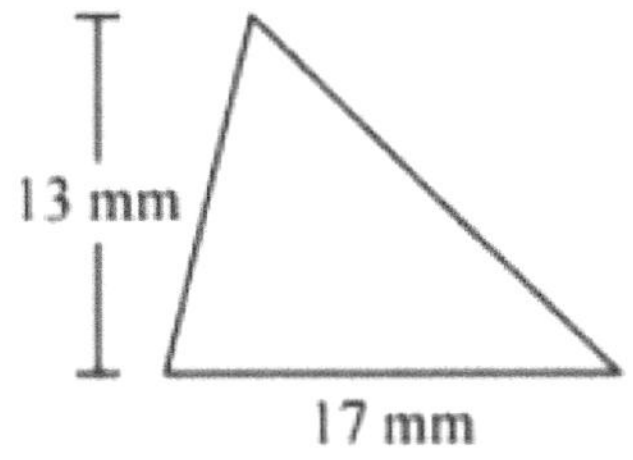

mm²

12. Calculate the area of the triangle shown. Write the answer in the box given below.

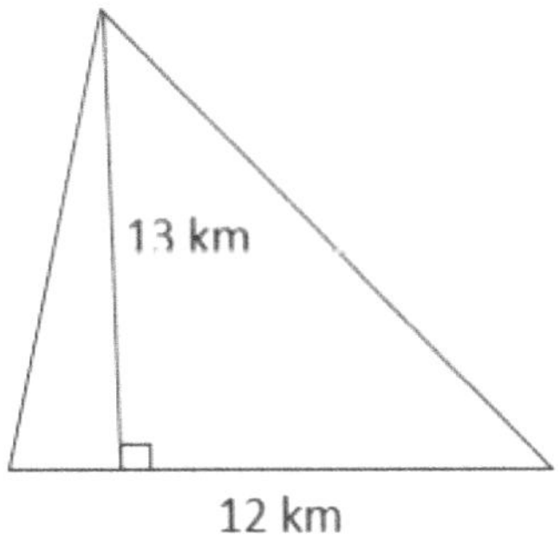

km²

13. **George and Maggie were completing their homework together. One problem asked to find the area of a triangle. Below are George's and Maggie's work.**

George's Work	Maggie's Work
$A = \frac{bh}{2}$	$A = \frac{bh}{2}$
$A = \frac{(3.2)(4.5)}{2}$	$A = \frac{(3.2)(3.8)}{2}$
$A = \frac{14.4}{2}$	$A = \frac{12.16}{2}$
$A = 7.2\ mm^2$	$A = 6.08\ mm^2$

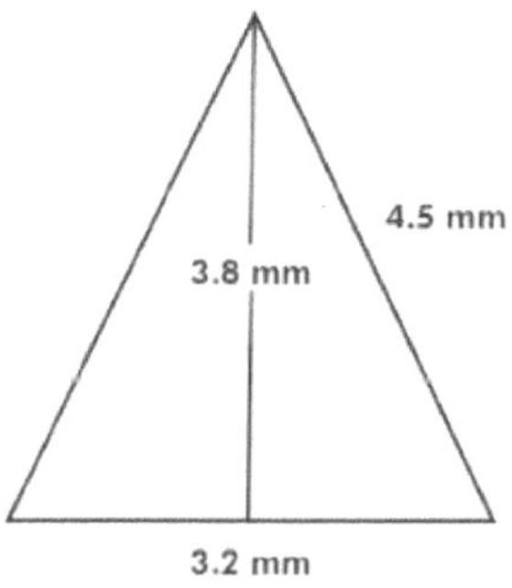

Whose answer is correct? Explain your reasoning.

Do NOT write your answers in this book. To open the answer sheet, scan the QR code or visit ***lumoslearning.com/a/6m032***

Chapter 4 → Lesson 2: Surface Area and Volume

1. How many rectangular faces would a trapezoidal prism have?

Ⓐ two
Ⓑ four
Ⓒ six
Ⓓ zero

2. Which of the following statements is true of a rhombus?

Ⓐ A rhombus is a parallelogram.
Ⓑ A rhombus is a quadrilateral.
Ⓒ A rhombus is equilateral.
Ⓓ All of the above are true.

3. A cube has a volume of 1,000 cm^3. What is its surface area?

Ⓐ 100 square cm
Ⓑ 60 square cm
Ⓒ 600 square cm
Ⓓ It cannot be determined.

4. A solid figure is casting a square shadow. The figure could <u>not</u> be a __________.

Ⓐ rectangular prism
Ⓑ cylinder
Ⓒ pentagonal pyramid
Ⓓ hexagonal prism

5. Which of the following solid figures has the most flat surfaces?

Ⓐ a cube
Ⓑ a triangular prism
Ⓒ a hexagonal prism
Ⓓ a pentagonal pyramid

6. **Calculate the surface area of the box shown below.**

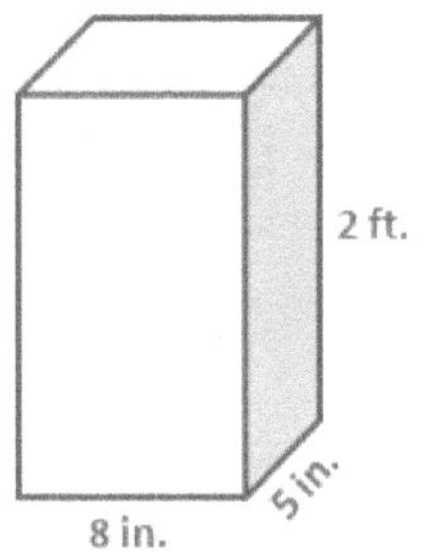

Ⓐ 132 square inches
Ⓑ 80 square inches
Ⓒ 704 square inches
Ⓓ 352 square inches

7. **Complete the following statement.**
A hexagon must have __________.

Ⓐ 6 sides and 6 angles
Ⓑ 8 sides and 8 angles
Ⓒ 10 sides and 10 angles
Ⓓ 7 sides and 7 angles

8. **A single marble tile measures 25 cm by 20 cm. How many tiles will be required to cover a floor with dimensions 2 meters by 3 meters?**

Ⓐ 320 tiles
Ⓑ 240 tiles
Ⓒ 180 tiles
Ⓓ 120 tiles

9. **Calculate the volume of the box shown below.**

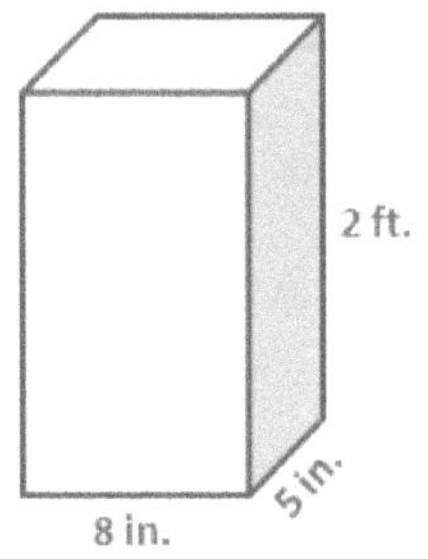

Ⓐ 80 cubic inches
Ⓑ 960 cubic inches
Ⓒ 800 cubic inches
Ⓓ None of the above

10. What is the surface area of a rectangular box with dimensions 4 cm, 6 cm, and 10 cm?

Ⓐ 248 square cm
Ⓑ 240 square cm
Ⓒ 124 square cm
Ⓓ 224 square cm

11. What is the volume of a rectangular prism that has a length of 10 cm, a width of 5 cm, and a height of 2 cm? Circle the correct answer choice.

Ⓐ V = 50 cubic cm
Ⓑ V = 17 cubic cm
Ⓒ V = 10 cubic cm
Ⓓ V = 100 cubic cm

12. Determine the volume of the prism shown. Write your answer in the box given below.

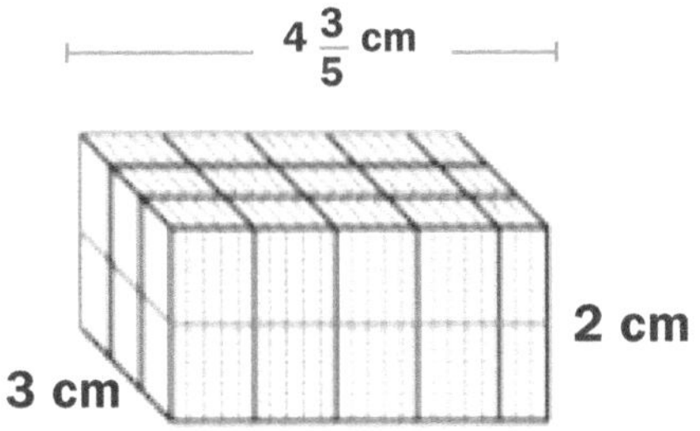

[ ] cm^3

Do NOT write your answers in this book. To open the answer sheet, scan the QR code or visit ***lumoslearning.com/a/6m033***

Chapter 4 → Lesson 3: Coordinate Geometry

1. The points A (0, 0), B (5, 0), C (6, 2), D (5, 5), and E (0, 5) are plotted in a coordinate grid. Describe the angles in pentagon ABCDE.

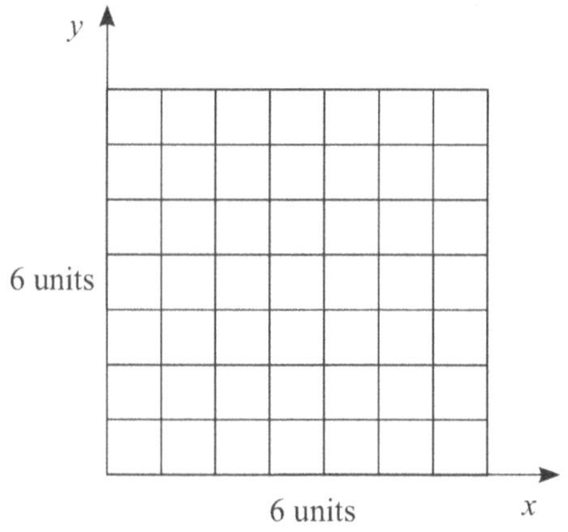

Ⓐ 2 right angles and 3 obtuse angles
Ⓑ 2 right angles, 1 obtuse angle, and 2 acute angles
Ⓒ 3 right angles and 2 obtuse angles
Ⓓ 2 right angles, 2 acute angles, and 1 obtuse angle

2. The corners of a shape are located at (1,2), (5,2), (2,3) and (4,3) in a coordinate grid. What type of shape is it?

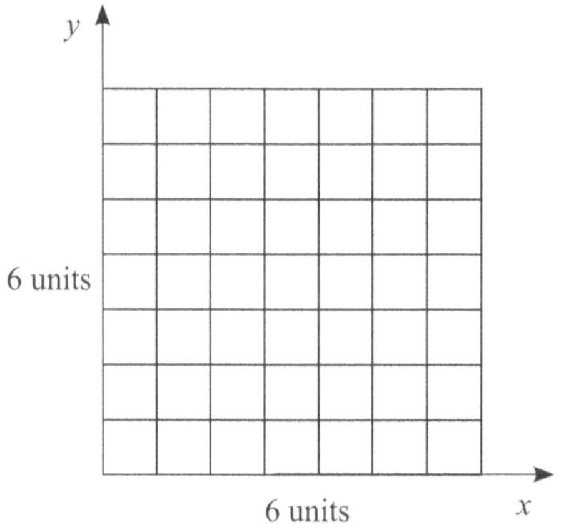

Ⓐ square
Ⓑ parallelogram
Ⓒ rhombus
Ⓓ trapezoid

3. **Which of the following graphs shows a 180-degree clockwise rotation about the origin?**

Ⓐ 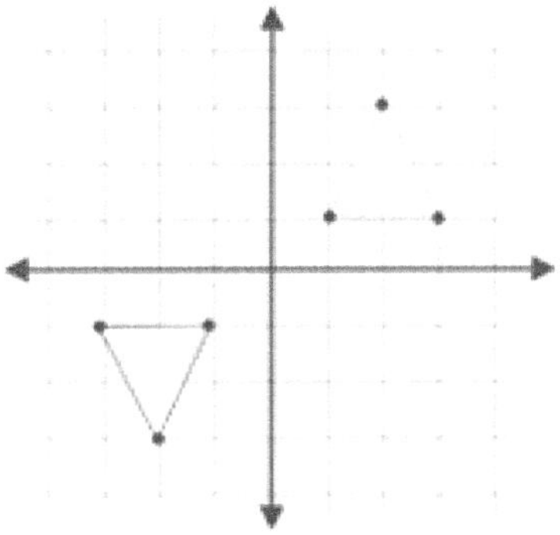

Ⓑ 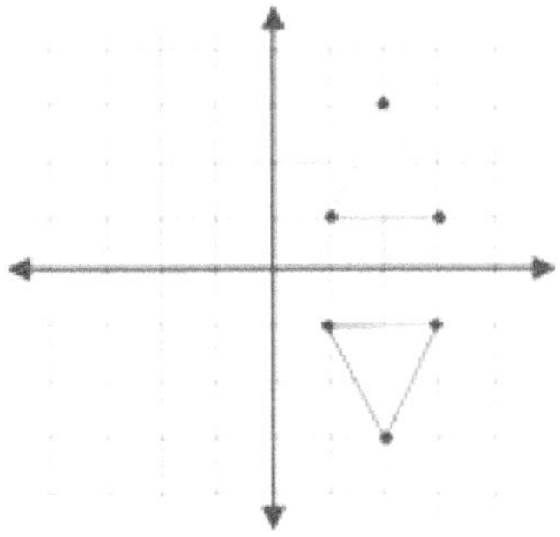

Ⓒ 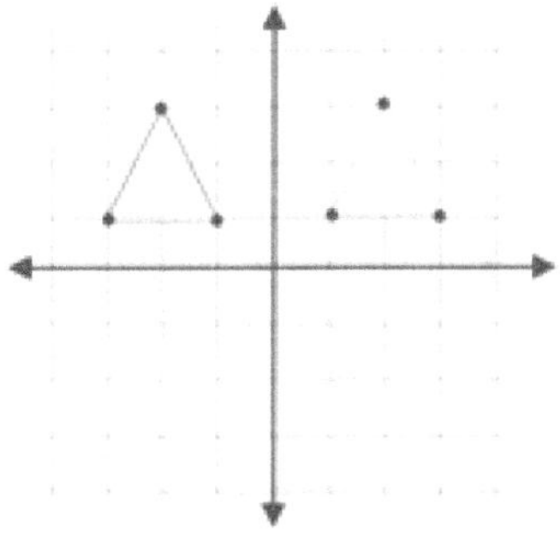

Ⓓ 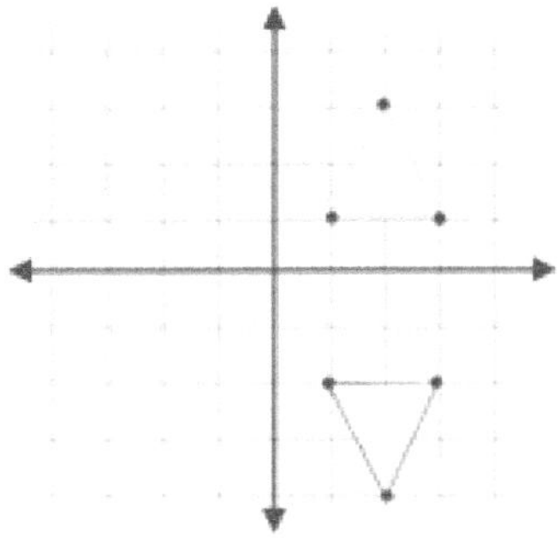

4. **Identify tx = 2y + 3, when y = 2.**

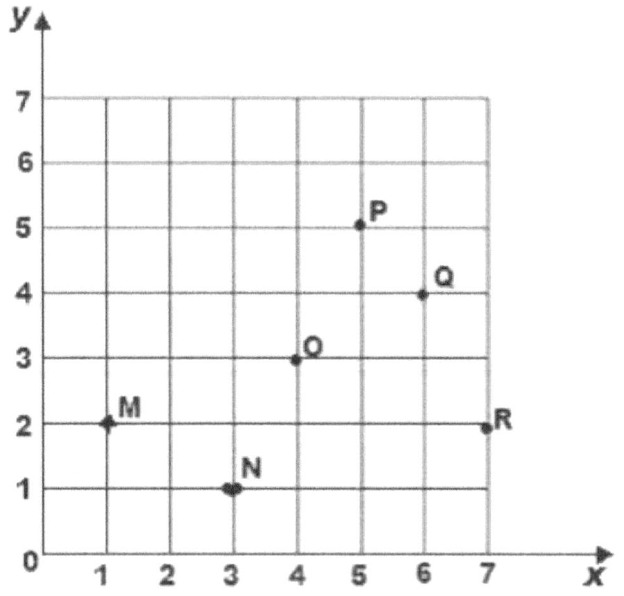

Ⓐ Point M
Ⓑ Point N
Ⓒ Point R
Ⓓ Point Q

5. **Which figure is formed when you draw straight line segments between the following points (in the order they are listed)? (2, −13), (2, 1), (8, 1), (8, 5), (2, 5), (2, 10), (−2, 10), (−2, 5), (−8, 5), (−8, 1), (−2, 1), (−2, −13)**

Ⓐ star
Ⓑ cross
Ⓒ heart
Ⓓ boat

6. **You are looking for a point on the line: y = 10 − 2x. You know that x = −1. What does y equal?**

Ⓐ 10
Ⓑ 8
Ⓒ 12
Ⓓ −2

7. **Assume a function has the rule y = 2x. Which grid shows the ordered pair formed when x = 3?**

Ⓐ

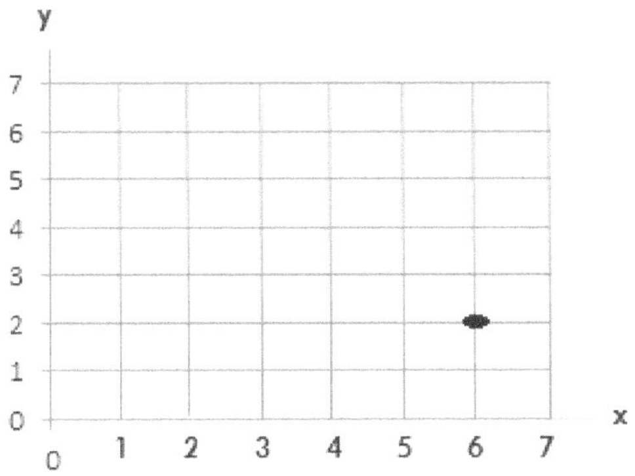

Ⓑ

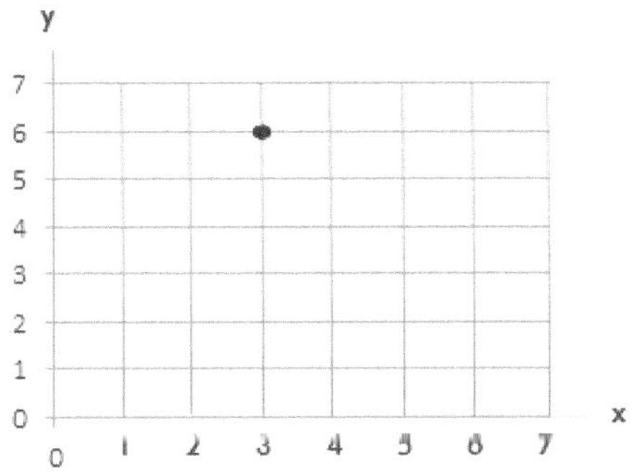

Ⓒ

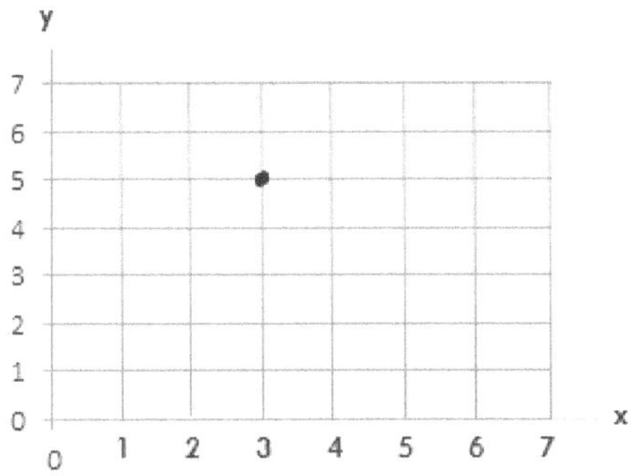

Ⓓ None of the above

8. **Which equation matches this graph?**

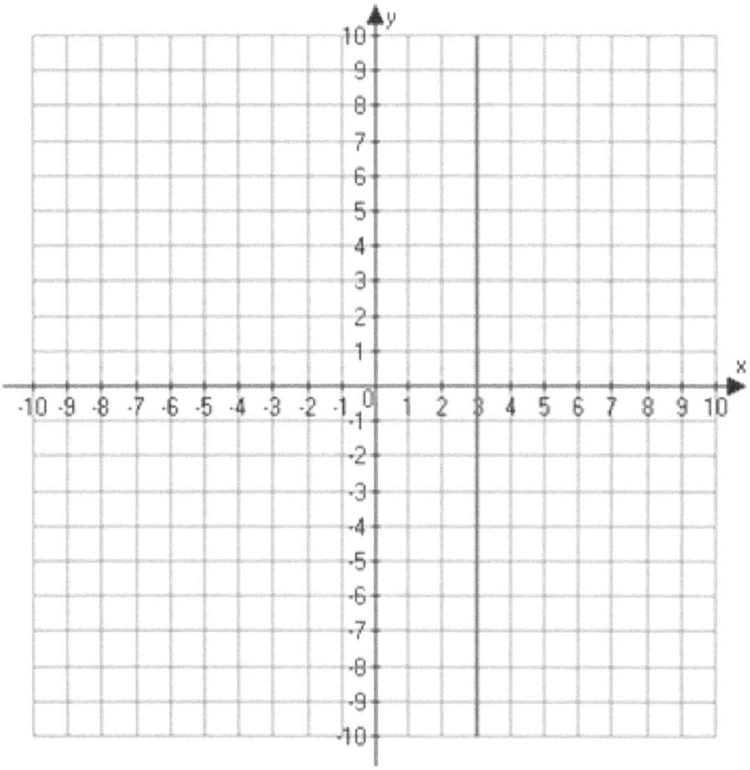

Ⓐ y = 3
Ⓑ x = 3
Ⓒ y = 0
Ⓓ x = −3

9. **What ordered pair would fit in this equation?**

y = x − 3

Ⓐ (4, 0)
Ⓑ (0, 4)
Ⓒ (4, 1)
Ⓓ (1, 4)

10. **The upper left region of the coordinate plane is Quadrant ____________**

Ⓐ IV
Ⓑ I
Ⓒ II
Ⓓ III

11. What is the area of this polygon? Write your answer in the box given below.

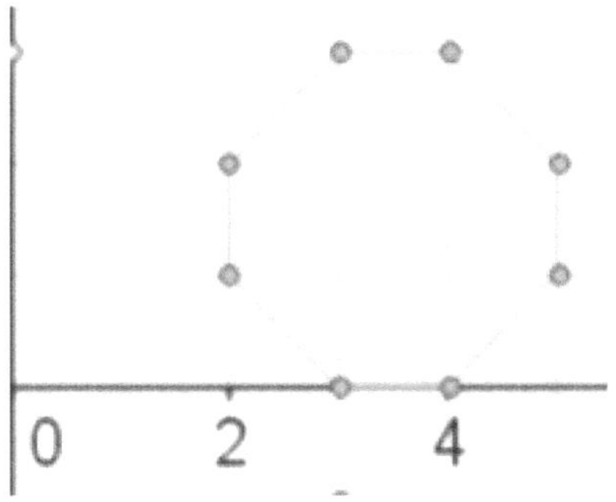

square units

12. What is the area of the rectangle. Write your answer in the box given below.

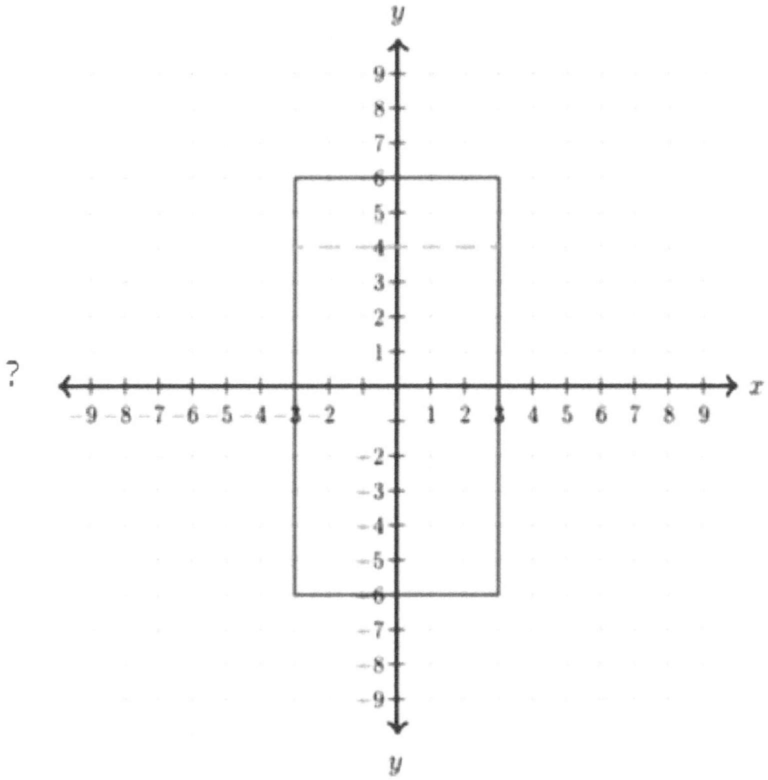

square units

Do NOT write your answers in this book. To open the answer sheet, scan the QR code or visit ***lumoslearning.com/a/6m034***

Chapter 4 → Lesson 4: Nets

1. Identify the solid given its net:

Ⓐ Rectangular prism
Ⓑ Cube
Ⓒ Triangular prism
Ⓓ Sphere

2. Identify the solid given its net:

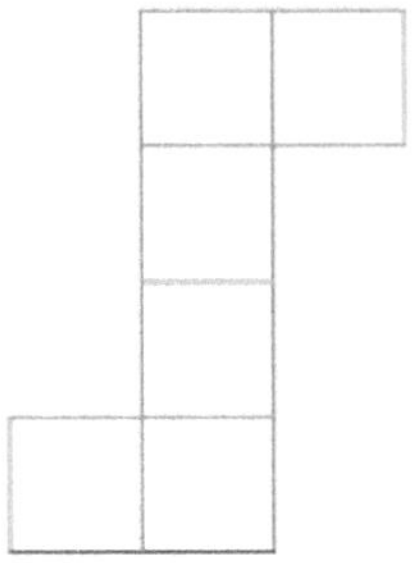

Ⓐ Cube
Ⓑ Sphere
Ⓒ Rectangular prism
Ⓓ Square pyramid

3. **Identify the solid given its net:**

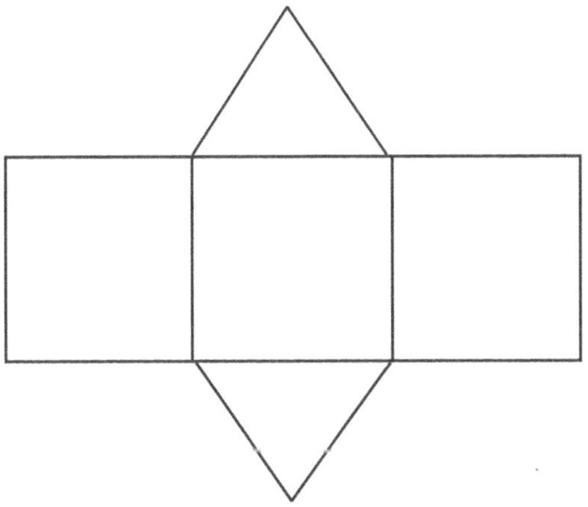

Ⓐ Rectangular prism
Ⓑ Cube
Ⓒ Triangular prism
Ⓓ Sphere

4. **Identify the solid given its net:**

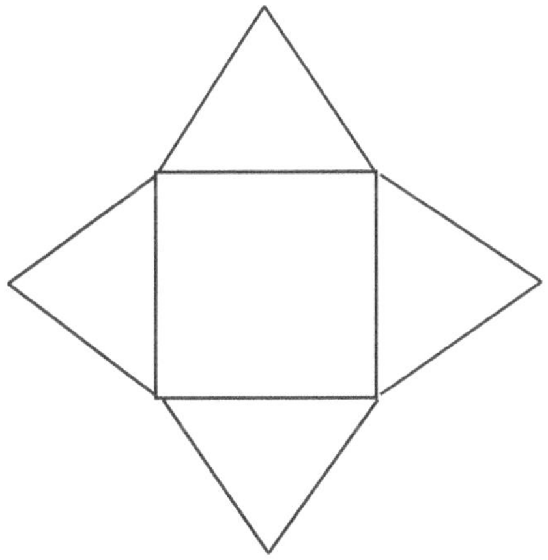

Ⓐ Cube
Ⓑ Sphere
Ⓒ Rectangular prism
Ⓓ Square pyramid

5. **Identify the solid given its net:**

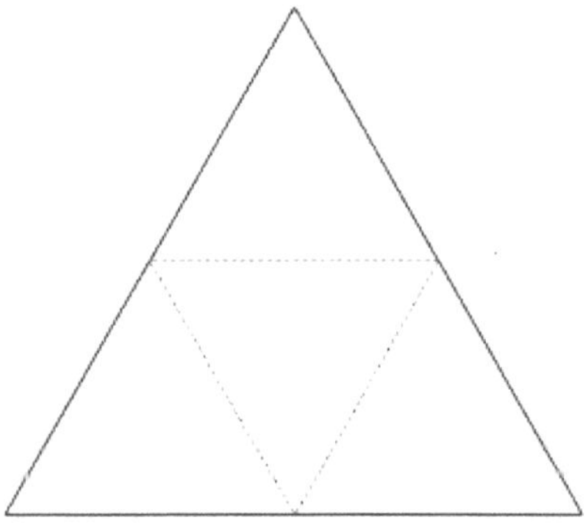

Ⓐ Cube
Ⓑ Sphere
Ⓒ Rectangular prism
Ⓓ Triangular pyramid

6. **Identify the solid given its net:**

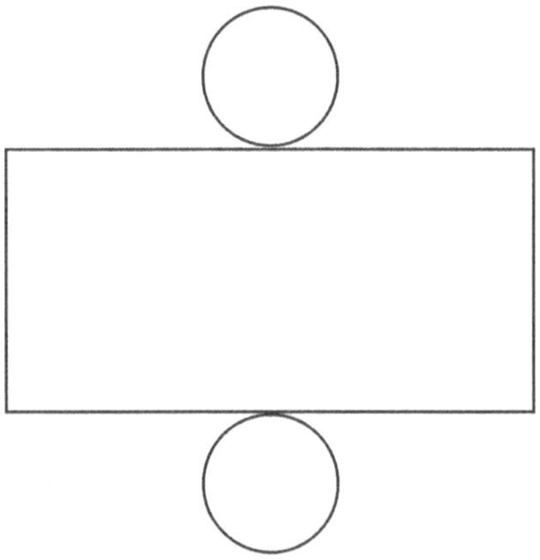

Ⓐ Cube
Ⓑ Sphere
Ⓒ Cylinder
Ⓓ Square pyramid

7. **Identify the solid given its net:**

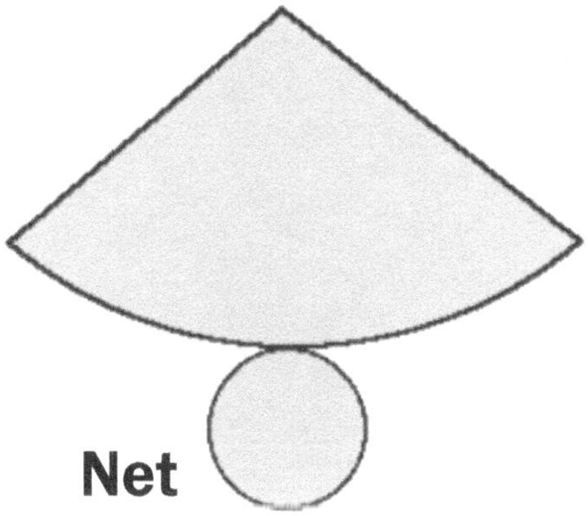

Ⓐ Cube
Ⓑ Cone
Ⓒ Cylinder
Ⓓ Square pyramid

8. **The diagram below represents the net of a solid figure. Find the surface area given L = 4 in, W = 4 in, H = 12 in.**

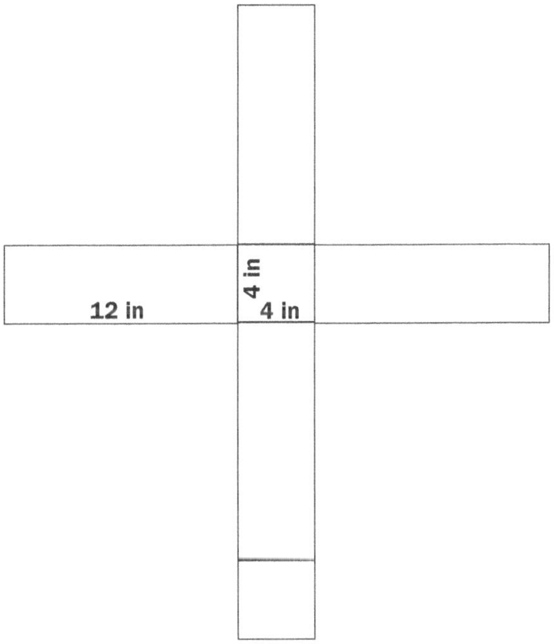

Ⓐ 228 square ft.
Ⓑ 224 square in.
Ⓒ 448 square in.
Ⓓ 112 square in.

9. **Identify the solid, given its net:**

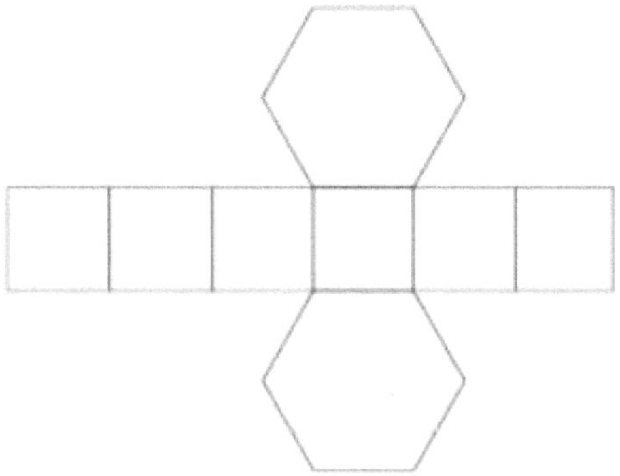

Ⓐ Sphere
Ⓑ Cone
Ⓒ Hexagonal prism
Ⓓ Hexagonal pyramid

10. **Identify the number of faces, edges and vertices in a rectangular pyramid.**

Ⓐ Faces = 5, Vertices = 6, Edges = 9
Ⓑ Faces = 4, Vertices = 4, Edges = 6
Ⓒ Faces = 6, Vertices = 8, Edges = 12
Ⓓ Faces = 5, Vertices = 5, Edges = 8

11. **Use a net or formula to find the surface area of the figure. The length is 5, the width is 4 and the height is 2. Write your answer in the box given below.**

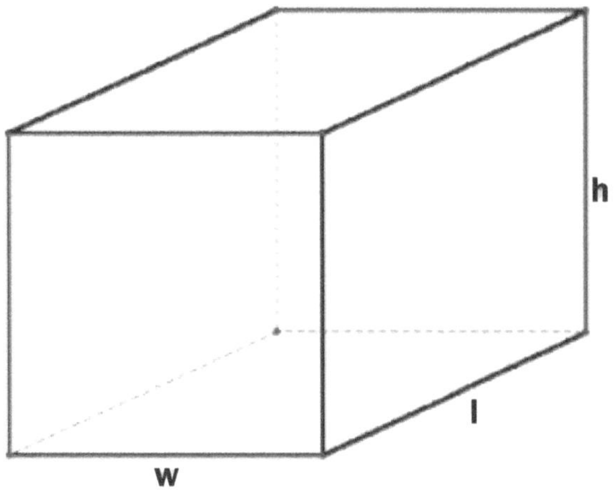

12. **Use a net or formula to find the surface area of the figure. Length is 2, width is 3, and height is 4. Circle the correct answer choice.**

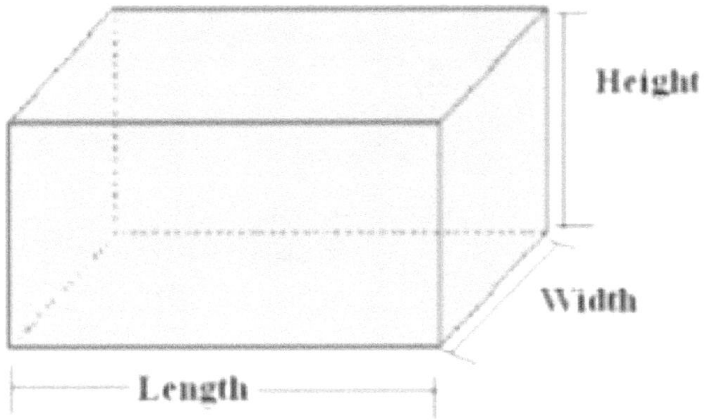

Ⓐ 52 square units
Ⓑ 42 square units
Ⓒ 32 square units
Ⓓ 22 square units

13. **Identify the number of faces, edges and vertices in a hexagonal prism. Circle the correct answer choice.**

Ⓐ Faces = 8, Vertices = 10, Edges = 16
Ⓑ Faces = 8, Vertices = 12, Edges = 20
Ⓒ Faces = 8, Vertices = 12, Edges = 18
Ⓓ Faces = 8, Vertices = 10, Edges = 18

End of Geometry

Chapter 5

Statistics & Probability

Do NOT write your answers in this book. To open the answer sheet, scan the QR code or visit ***lumoslearning.com/a/6m035***

Lesson 1: Statistical Questions

1. The chart below shows the participation of a sixth grade class in its school's music activities. Each student was allowed to pick one music activity.

How many boys are in the sixth grade?

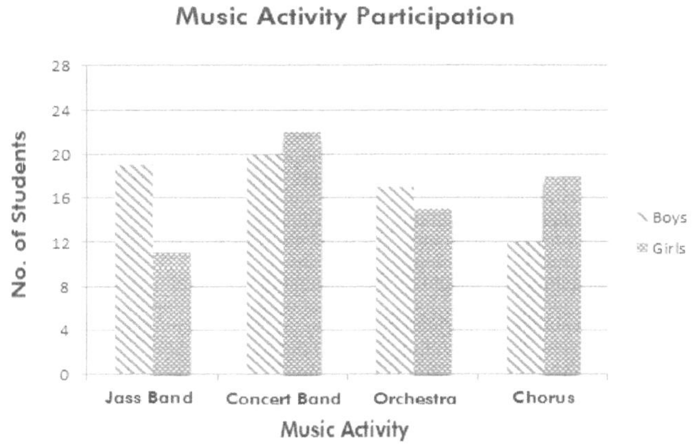

Ⓐ 86
Ⓑ 68
Ⓒ 42
Ⓓ 12

2. A student wanted to know what the sixth grade girls' favorite song was. What would be the best way to conduct a survey?

Ⓐ Do an Internet search of favorite songs of young girls.
Ⓑ Survey the sixth grade boys.
Ⓒ Survey the sixth grade girls.
Ⓓ Conduct a survey at the mall.

3. Emily wanted to know what the range of daily temperatures this past week was. What would be the best way of conducting her survey?

Ⓐ Take the temperature once during the week.
Ⓑ Take the temperature only in the morning.
Ⓒ Take the temperature only in the evening.
Ⓓ Take the temperature two times a day, at the warmest and coolest times of the day.

4. Goldie wants to find out how many presents children get for their birthdays. She surveys 11 families in the same neighborhood to find out how many presents their children get. Did Goldie get a representative sample?

Ⓐ No because she did not ask the right questions.
Ⓑ No because she asked families in the same neighborhood who most likely have similar income levels.
Ⓒ Yes because she asked families in the same neighborhood who most likely have similar income levels.
Ⓓ Yes because she asked the right questions.

5. Roberta is an arborist. She is studying maple trees in a specific area. Roberta wants to show a class the difference in the heights of the trees so that they can compare them. What type of graph would be best for that?

Ⓐ Line graph
Ⓑ Picture graph
Ⓒ Circle graph
Ⓓ Bar graph

6. Derek spends an average of 37 minutes a weekday on homework. He wants to know how much time other students in fourth grade spend on homework so he asks only students in his class. Will Derek's survey be biased?

Ⓐ Yes because he is asking fourth graders.
Ⓑ No because he is asking fourth graders.
Ⓒ Yes because students in his class have the same amount of homework as he does.
Ⓓ No because students in his class have the same amount of homework as he does.

7. There are five cities in New York State who compete for the title of "The Snowiest City." A survey is done to find the average snowfall. The average snowfall of all 5 cities is 112.6 inches. Cities 1 and 2 both get 116 inches of snow. City 3 gets 110.4 inches of snow. City 4 gets 119.6 inches of snow. City 5 gets 101 inches of snow. Which graph would best represent this information?

Ⓐ Picture graph
Ⓑ Circle graph
Ⓒ Bar graph
Ⓓ Line graph

8. Brooke is doing a survey to find out what percentage of time families spend together doing activities. She collects all of the data about the time spent together doing those different activities and wants to put it into a graph. Which graph would be best?

Ⓐ Circle graph
Ⓑ Picture graph
Ⓒ Line graph
Ⓓ Bar graph

9. Peter and Paul are playing cards. They each randomly select 12 cards to start the game. Are the cards that they each selected a biased sample?

Ⓐ No because they were randomly selected.
Ⓑ Yes because they were randomly selected.
Ⓒ Yes because they are not at least 10% of the deck.
Ⓓ No because they are not at least 10% of the deck.

10. Cara likes to go running. She runs 4 days a week. On Monday, she runs 7 miles. On Wednesday, she runs 3.4 miles. On Friday, she runs 5 miles. On Saturday she runs 7.4 miles. Cara wants to put her data into a graph so that she can visually see the fluctuation in the miles she runs. What graph would be best?

Ⓐ Circle graph
Ⓑ Bar graph
Ⓒ Line graph
Ⓓ Picture graph

11. Select the questions that qualify as statistical questions. Choose all that apply.

Ⓐ How many letters are in my last name?
Ⓑ How many letters are in the last names of the students in my 6th grade class?
Ⓒ What are the colors of the shoes worn by the students in my school?
Ⓓ What are the heart rates of the students in a 6th grade class?
Ⓔ How many hours of sleep per night do 6th graders usually get when they have school the next day?

12. Match the correct data plot for each question. Write the letters (in capitals) in the box.

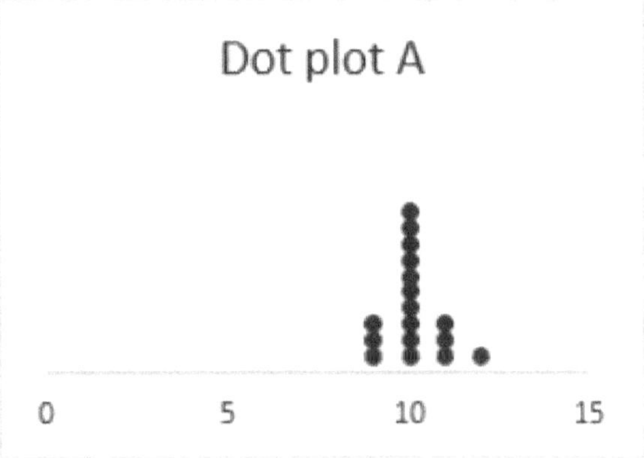

Question	Dot Plot ___
1. What are the ages of 4th graders in our school?	
2. What are the heights of the players on the 8th grade boys' basketball team?	
3. How many hours do 6th graders in our class watch TV on a school night?	
4. How many different languages do students in our class speak?	

Do NOT write your answers in this book. To open the answer sheet, scan the QR code or visit ***lumoslearning.com/a/6m036***

Chapter 5 → Lesson 2: Distribution

1. If the total sales for socks was $60, what is the best estimate for the total sales of pants?

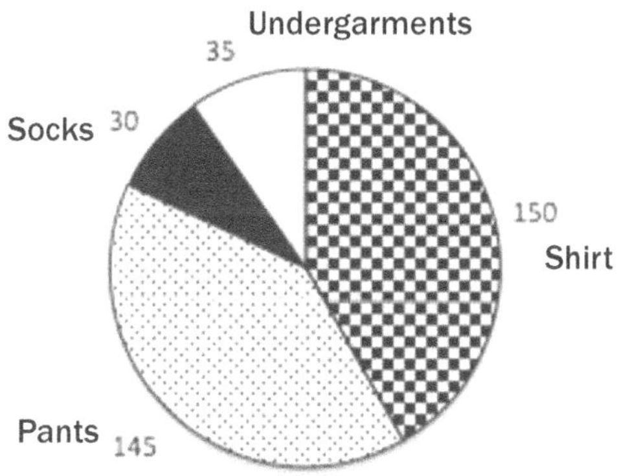

Note: Data represent the angles for each category

Ⓐ $200
Ⓑ $300
Ⓒ $60
Ⓓ $1,000

2. The sixth graders at Kilmer Middle School can choose to participate in one of the four music activities offered. The number of students participating in each activity is shown in the bar graph below. Use the information shown to respond to the following: How many sixth graders are in the Jazz Band?

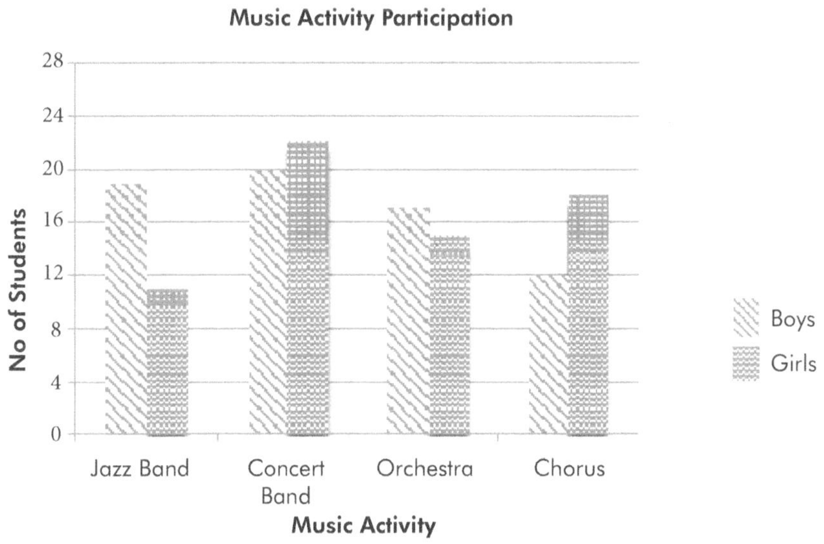

Ⓐ 19 sixth graders
Ⓑ 20 sixth graders
Ⓒ 30 sixth graders
Ⓓ 25 sixth graders

3. **A .J. has downloaded 400 songs onto his computer. The songs are from a variety of genres. The circle graph below shows the breakdown (by genre) of his collection. Use the information shown to respond to the following: About how many more R + B songs than rock songs has A.J. downloaded?**

A.J.'s Music Collections

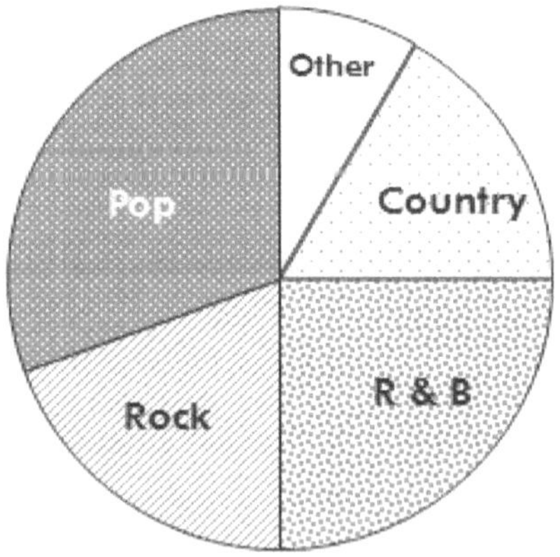

Ⓐ 20 more songs
Ⓑ 30 more songs
Ⓒ 50 more songs
Ⓓ 75 more songs

4. **Colleen has to travel for work. In one week, she traveled all five work days. The shortest distance she traveled was 63 miles. The range of miles that she traveled was 98 miles. What is the longest distance that Colleen traveled for work in one week?**

Ⓐ 161 miles
Ⓑ 98 miles
Ⓒ 33 miles
Ⓓ It cannot be determined.

5. **Bob is a mailman. He delivers a lot of letters every day. His mail bag will only hold so many letters. The most letters that Bob has ever delivered in a day is 8,476. The range of the number of letters that Bob has ever delivered is 6,930. What is the least number of letters that Bob has ever delivered in one day?**

Ⓐ 15,406 letters
Ⓑ 1,546 letters
Ⓒ 1,556 letters
Ⓓ It cannot be determined.

6. Fred goes with Katie to the art sale to try to find a gift for his mother. There are 66 pieces of art for sale. The median price is $280. Fred has $280 to spend. What is the minimum number of art pieces Fred can choose from?

Ⓐ 16
Ⓑ 33
Ⓒ 65
Ⓓ 22

7. Joann loves to watch the birds outside of her window. There is a robin that comes back every year to build a nest and lay her eggs. Joann keeps track of how many eggs the robin lays. The first year Joann keeps track, the robin lays 9 eggs and that is the least number of eggs she lays. The range of eggs that the robin has laid is one third the number of eggs from the first year. What are the most eggs the robin has ever laid?

Ⓐ 21 eggs
Ⓑ 12 eggs
Ⓒ 14 eggs
Ⓓ 27 eggs

8. Matt loves to watch movies. He has a movie collection that has 285 movies in it. The length of time of all of his movies range from 48 minutes to 3 hours and 26 minutes. What is the range of the length of time of Matt's movies?

Ⓐ 3 hours and 38 minutes
Ⓑ 1 hour and 58 minutes
Ⓒ 2 hours and 58 minutes
Ⓓ 2 hours and 38 minutes

9. Jessica is a photographer. She is putting together photo albums to show off her photographs. The first photo album has 50 pictures in it. The second photo album has 119 pictures in it. The third photo album has 174 pictures in it. The fourth photo album has 72 pictures in it. The fifth photo album has 61 pictures in it. What is the range?

Ⓐ 102
Ⓑ 174
Ⓒ 124
Ⓓ 113

10. Sam is a shrimp fisherman. In one month, Sam gets a maximum of 280 pounds of shrimp. The minimum that Sam gets is half of his maximum. What is the range in the pounds of shrimp that Sam gets?

Ⓐ 120
Ⓑ 280
Ⓒ 100
Ⓓ 140

11. Wrapping paper rolls are sold in packages of 2, 4, or 6. The following prices apply to the packages:

Package of 2 rolls —— $1.50
Package of 4 rolls —— $2.50
Package of 6 rolls —— $3.50

Lisa needs to buy exactly 12 rolls of wrapping paper for a service project.
Below are a few of the possible combinations. Select the value that correctly represents the missing number in the combinations listed on the table.

	2	4	6
2, 2, 2, 2, 2, **a** or 6 ($1.50) = $9.00: a =	○	○	○
2, 2, 2, **b**, 4 or 4($1.50)) + $2.50 = $8.50: b =	○	○	○
2, 2, **c**, 4, or 2($1.50) + 2($2.50) = $8.00: c =	○	○	○

12. The height of the four tallest downtown buildings of three different cities is given in the table below. Calculate the mean height of each city's buildings and match it with the correct answer from the bottom table.

City A Buildings	Height (m)
1	23
2	32
3	35
4	45

City B Buildings	Height (m)
1	15
2	22
3	28
4	25

City C Buildings	Height (m)
1	15
2	30
3	35
4	36

	22.50 m	29 m	23.20 m	33.75 m	18 m
City A Buildings	○	○	○	○	○
City B Buildings	○	○	○	○	○
City C Buildings	○	○	○	○	○

Do NOT write your answers in this book. To open the answer sheet, scan the QR code or visit ***lumoslearning.com/a/6m037***

Chapter 5 → Lesson 3: Central Tendency

1 **Jason was conducting a scientific experiment using bean plants. He measured the height (in centimeters) of each plant after three weeks. These were his measurements (in cm): 12, 15, 11, 17, 19, 21, 13, 11, 16**

What is the average (mean) height? What is the median height?

Ⓐ Mean = 15 cm, Median = 19 cm
Ⓑ Mean = 15 cm, Median = 15 cm
Ⓒ Mean = 19 cm, Median = 15 cm
Ⓓ Mean = 15 cm, Median = 13 cm

2. **Stacy has 60 pairs of shoes. She has shoes that have a heel height of between 1 inch and 4 inches. Stacy has 20 pairs of shoes that have a 1 inch heel, 15 pairs of shoes that have a 2 inch heel and 20 pairs of shoes that have a 3 inch heel height. Remaining shoes have 4 inch heel height. What is the average heel height of all 60 pairs of Stacy's shoes?**

Ⓐ 1.3 inches
Ⓑ 1.5 inches
Ⓒ 2 inches
Ⓓ 2.2 inches

3. **What is the median of the following set of numbers?**
{16, −10, 13, −8, −1, 5, 7, 10}
Ⓐ −8
Ⓑ −1
Ⓒ 4
Ⓓ 6

4. **Given the following set of data, is the median or the mode larger?**
{5, −10, 14, 6, 8, −2, 11, 3, 6}
Ⓐ The mode
Ⓑ The median
Ⓒ They are the same
Ⓓ You cannot figure it out

5. A = { 10, 15, 2, 14, 19, 25, 0 }

Which of the following numbers, if added to Set A, would have the greatest effect on its median?

Ⓐ 14
Ⓑ 50
Ⓒ 5
Ⓓ 15

6. What is the mean, median and mode for the following data?

{-7, 18, 29, 4, -3, 11, 22}

Ⓐ Mean = 10
Median = 18
Mode = -3

Ⓑ Mean = 11
Median = 11
Mode = none

Ⓒ Mean = 10.57
Median = 11
Mode = none

Ⓓ Mean = 10
Median = 18
Mode = -3

7. Amanda had the following numbers: 1,2,6

If she added the number 3 to the list...

Ⓐ the mean would increase
Ⓑ the mean would decrease
Ⓒ the median would increase
Ⓓ the median would decrease

8. A pizza shop sells the following ice cream treats:

Strawberry Gelato: $0.60
Chocolate Cone: $0.80
Lemon Ice: $0.45

What is the average sale price for the ice cream treats?

Ⓐ $0.62
Ⓑ $0.45
Ⓒ $0.60
Ⓓ $0.80

9. Marcel has 5 stamp collections. He wants to average 35 stamps per collection. So far, he has 28, 62, 12, and 44 stamps in each collection. How many stamps does he need to have in his 5th collection to average 35 stamps?

Ⓐ 30
Ⓑ 29
Ⓒ 28
Ⓓ 27

10. How much will the mean of the following set of numbers increase if the number 53 is added to it?
{67, 29, 40, −12, 88, −7, 11}

Ⓐ 53
Ⓑ 2.768
Ⓒ 2.571
Ⓓ 6.625

11. Find the mean of the following set.
{ 25.1, 19.6, 88.5, 0, -3, 19.8 }
Circle the correct answer.

Ⓐ 19.6
Ⓑ 26
Ⓒ 25
Ⓓ 19.7

12. Select the median number for each set of numbers.

	3	4	8
4, 2, 3, 6, 4, 9, 7	○	○	○
6, 3, 2, 8, 1, 3, 6	○	○	○
8, 9, 4, 8, 1, 10, 3	○	○	○

13. Select the median number for each set of numbers.

	11	15	18
12, 18, 11, 20, 20	○	○	○
14, 20, 18, 11, 15	○	○	○
8, 6, 15, 11, 18	○	○	○

Do NOT write your answers in this book. To open the answer sheet, scan the QR code or visit ***lumoslearning.com/a/6m038***

Chapter 5 → Lesson 4: Graphs & Charts

1. Which of the following graphs best represents the values in this table?

X	Y
1	1
2	3
3	1
4	3

Ⓐ
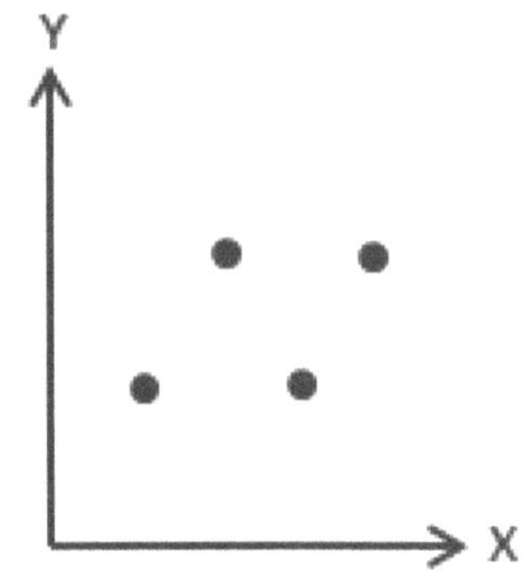

Ⓑ
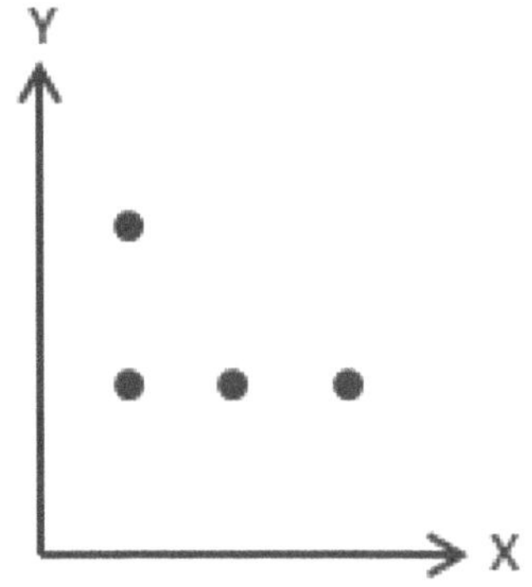

Ⓒ
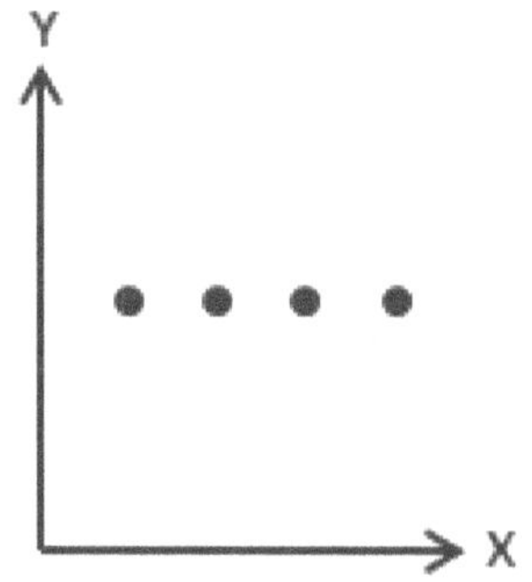

Ⓓ
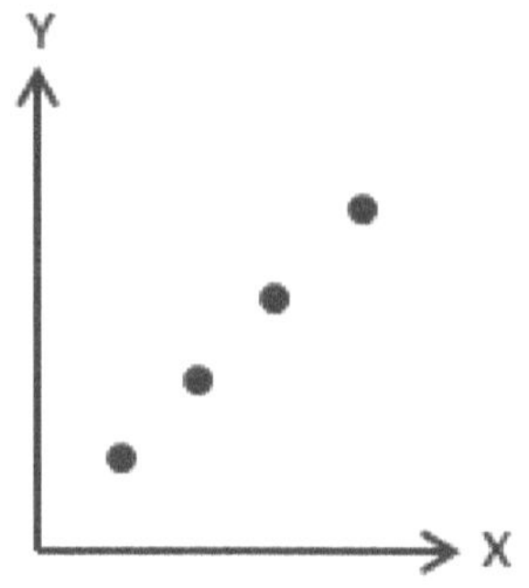

2. **The results of the class' most recent science test are displayed in this histogram. Use the results to answer the question. A "passing" score is 61 or higher.**

How many students passed the science test?

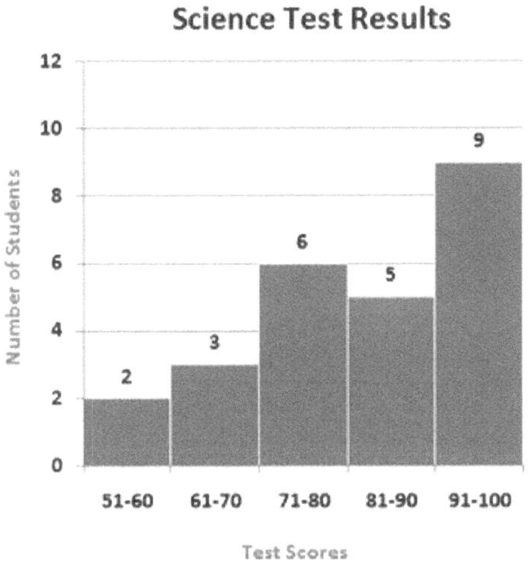

Ⓐ 3 students
Ⓑ 20 students
Ⓒ 23 students
Ⓓ 25 students

3. **The results of the class' most recent science test are displayed in this histogram. Use the results to answer the question. How many students scored a 90 or below?**

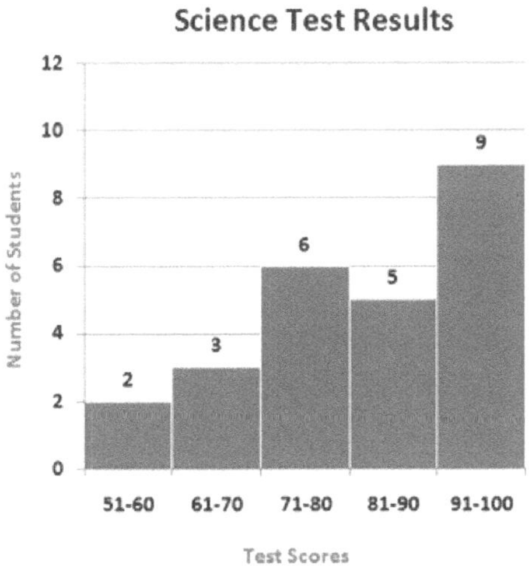

Ⓐ 5 students
Ⓑ 9 students
Ⓒ 11 students
Ⓓ 16 students

4. **As part of their weather unit, the students in Mr. Green's class prepared a line graph showing the high and low temperatures recorded each day during a one-week period. Use the graph to answer the question.**

On which day was the greatest range in temperature seen?

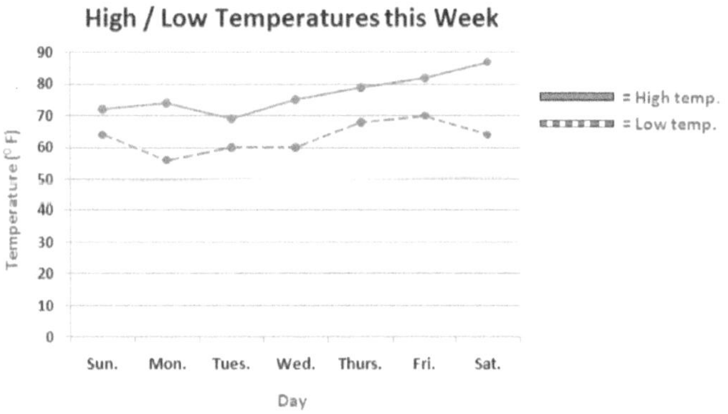

Ⓐ Sunday
Ⓑ Monday
Ⓒ Friday
Ⓓ Saturday

5. **The sixth graders at Kilmer Middle School can choose to participate in one of the four music activities offered. The number of students participating in each activity is shown in the bar graph below. Use the information shown to answer the question.**

There are 180 sixth graders in the school. About how many do not participate in one of the music activities?

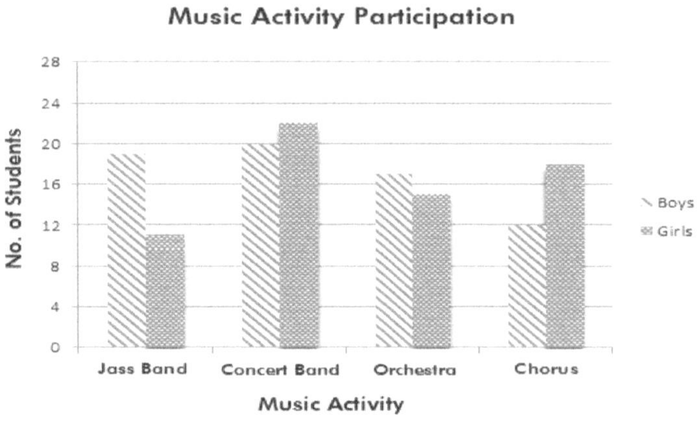

Ⓐ 66 students
Ⓑ 46 students
Ⓒ 25 students
Ⓓ 15 students

6. **A.J. has downloaded 400 songs onto his computer. The songs are from a variety of genres. The circle graph below shows the breakdown of his collection by genre. Use the information shown to answer the question.**

Which two genres together make up more than half of A.J.'s collection?

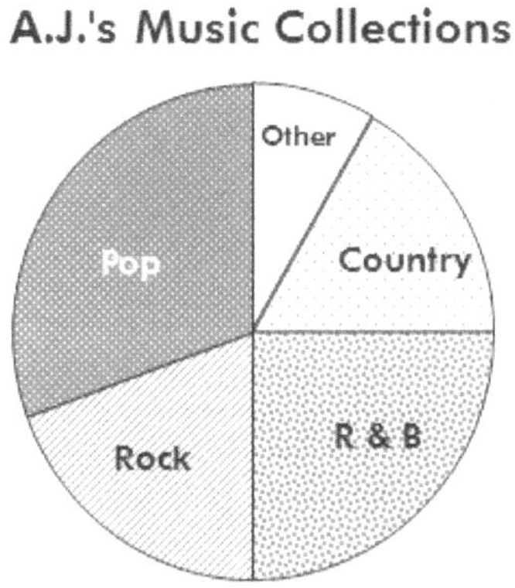

Ⓐ R + B and Country
Ⓑ Country and Rock
Ⓒ Rock and Pop
Ⓓ Pop and R + B

7. **The results of the class' most recent science test are displayed in this histogram. Use the results to answer the question.**

What percentage of the class scored an 81-90 on the test?

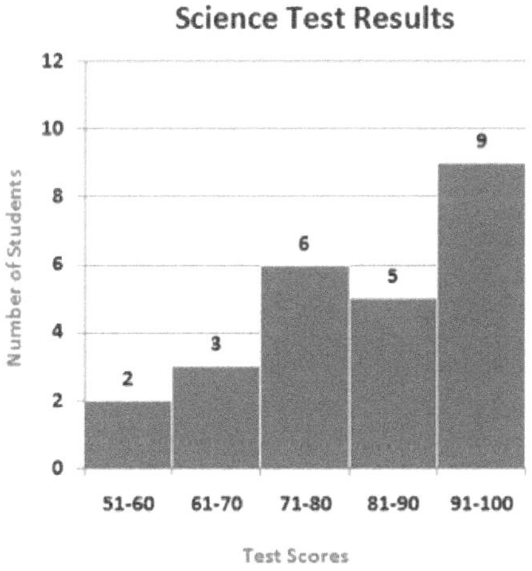

Ⓐ 5%
Ⓑ 20%
Ⓒ 25%
Ⓓ 30%

8. **How much of the graph do undergarments and socks make up together?**

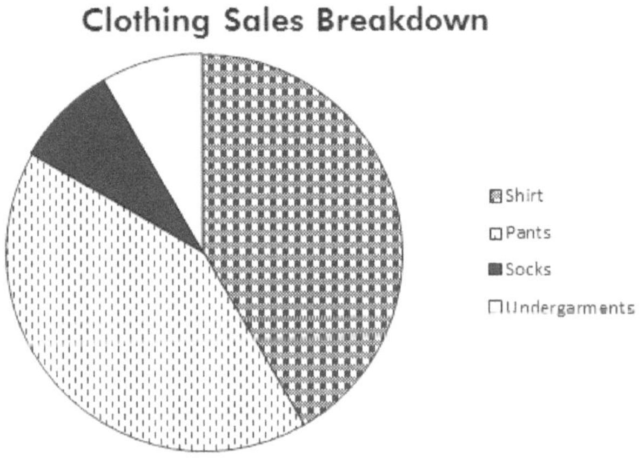

Ⓐ less than 5%
Ⓑ between 5% and 10%
Ⓒ between 10% and 25%
Ⓓ more than 25%

9. **As part of their weather unit, the students in Mr. Green's class prepared a line graph showing the high and low temperatures recorded each day during a one-week period. Use the graph to answer the question.**

What percentage of the days had a high temperature of 80 degrees or higher? Round to the nearest tenth.

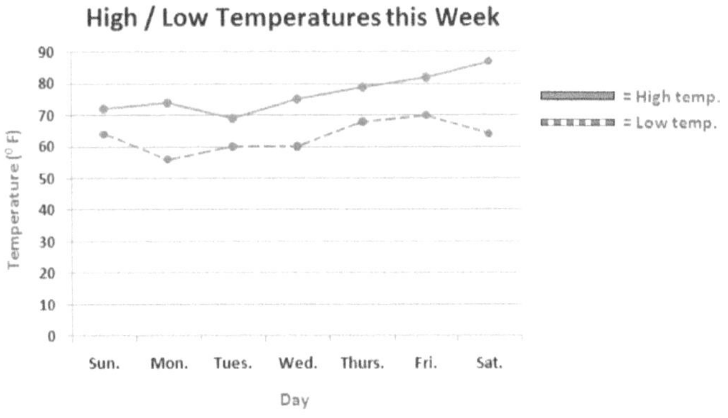

Ⓐ 14.3%
Ⓑ 42.9%
Ⓒ 28.6%
Ⓓ None of these

10. **The sixth graders at Kilmer Middle School can choose to participate in one of the four music activities offered. The number of students participating in each activity is shown in the bar graph below. Use the information shown to answer the question.**

There are 132 students who participate in music activities. What percentage of students who participate in music activities participate in chorus or jazz band?

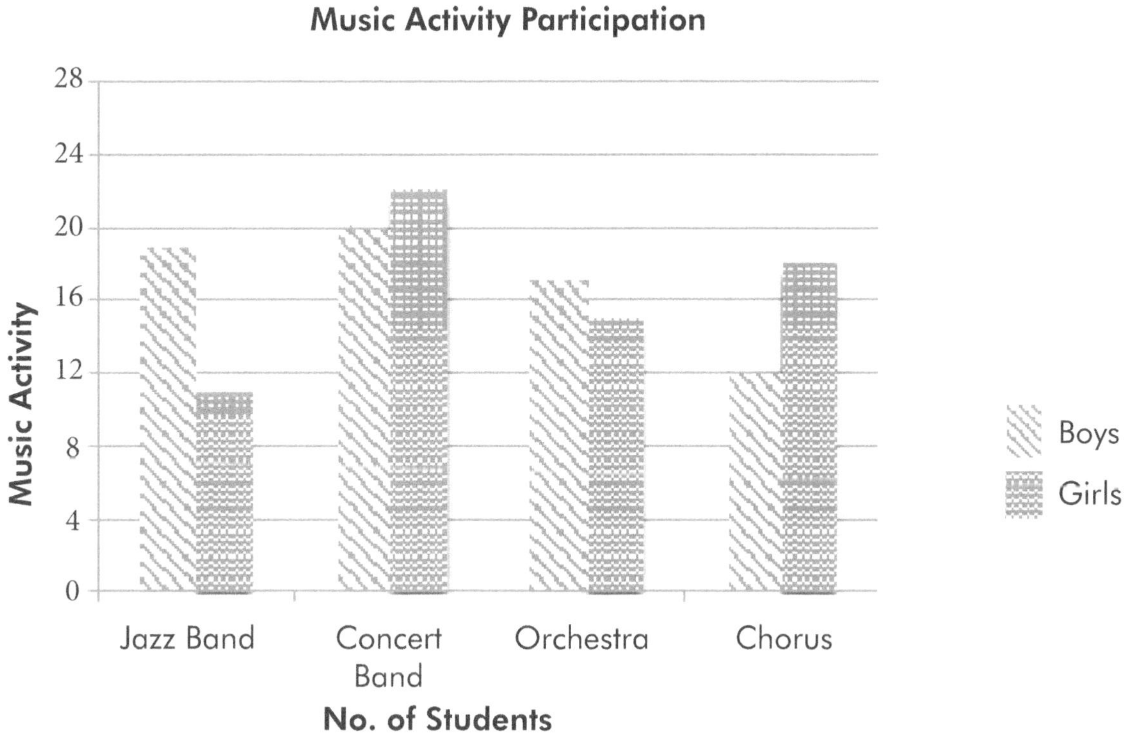

- Ⓐ 45%
- Ⓑ 38%
- Ⓒ 40%
- Ⓓ 50%

Questions 11 and 12 are based on the bar graph shown below.

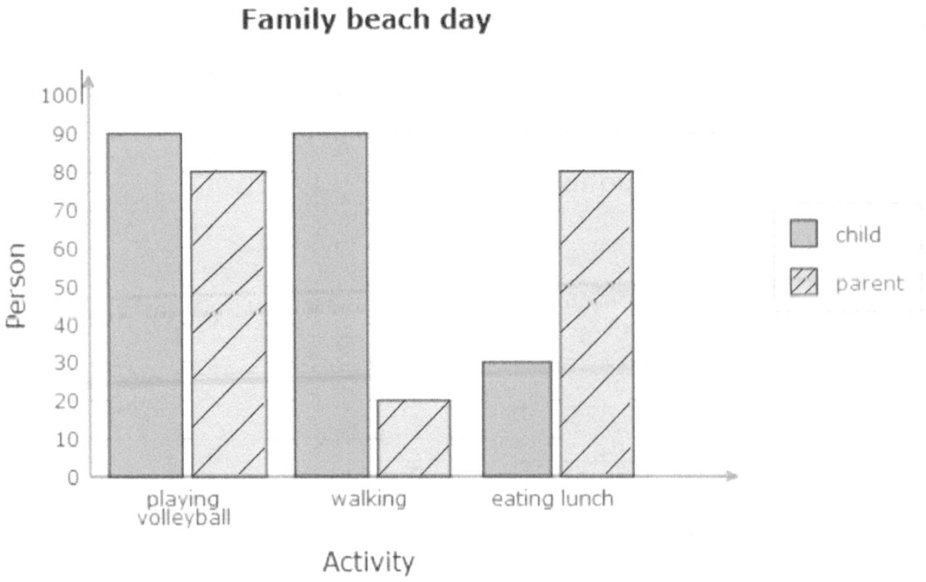

11. Which of the following statements are true? Select all that apply.

Ⓐ The same number of children participated in playing volleyball and walking
Ⓑ The same number of children participated in walking and eating lunch.
Ⓒ All of the children who played volleyball also ate lunch.
Ⓓ More children than parents ate lunch.
Ⓔ More children than parents played volleyball.
Ⓕ The same number of parents played volleyball as ate lunch.

12. What is the total number of parents and children who played volleyball and ate lunch? Write your answer in the box below.

Do NOT write your answers in this book. To open the answer sheet, scan the QR code or visit ***lumoslearning.com/a/6m039***

Chapter 5 → Lesson 5: Data Interpretation

1. **In the last 4 seasons, Luis scored 14, 18, 15, and 25 goals. How many goals does he need to score this year, to end the season with an overall average of 20 goals per season?**

   Ⓐ 29
   Ⓑ 27
   Ⓒ 30
   Ⓓ 28

2. **Stacy has 60 pairs of shoes. She has shoes that have a heel height of between 1 inch and 4 inches. Stacy wants to know which heel height she has the most of. Would she figure out the mode or mean?**

   Ⓐ Mean
   Ⓑ Mode
   Ⓒ Both
   Ⓓ Neither

3. **There are three ice cream stands within 15 miles and they are owned by Mr. Sno.**

Ice Cream Stand	Vanilla	Chocolate	Twist
A	15	22	10
B	24	8	14
C	20	16	13

   What percentage of the ice cream cones sold by Ice Cream Stand B were vanilla? Round your answer to the nearest whole number.

   Ⓐ 52%
   Ⓑ 24%
   Ⓒ 50%
   Ⓓ 25%

4. There are three ice cream stands within 15 miles and they are owned by Mr. Sno.

Ice Cream Stand	Vanilla	Chocolate	Twist
A	15	22	10
B	24	8	14
C	20	16	13

What percentage of the ice cream cones sold by Ice Cream Stand C were chocolate? Round your answer to the nearest whole number.

Ⓐ 30%
Ⓑ 32%
Ⓒ 33%
Ⓓ 62%

5. Alexander plays baseball. His batting averages for the games he played this year were recorded. What is the batting average he had the most often?

{.228, .316, .225, .333, .228, .125, .750, .500}

Ⓐ .228
Ⓑ .316
Ⓒ .750
Ⓓ .333

6. How much will the mean increase by when the number 17 is added to the set? Round your numbers to the nearest tenth.

{5, -10, 14, 6, 8, -2, 11, 3, 6}

Ⓐ 1.5
Ⓑ 1
Ⓒ 1.2
Ⓓ 1.7

7. Andrea plays the violin. She is practicing songs for her concert and she times how long it takes her to play each song. She wants to put the song with the middle number of minutes at the beginning of her concert. How many minutes is the song that she will play first?

{13, 8, 4, 16, 3, 9, 11}

Ⓐ 13 minutes
Ⓑ 11 minutes
Ⓒ 8 minutes
Ⓓ 9 minutes

8. There are three ice cream stands within 15 miles and they are owned by Mr. Sno.

Ice Cream Stand	Vanilla	Chocolate	Twist
A	15	22	10
B	20	8	14
C	24	16	13

What percentage of the total ice cream cones sold were twist? Round your answer to the nearest whole number.

Ⓐ 26%
Ⓑ 24%
Ⓒ 62%
Ⓓ 14%

9. Amy is trying to find out the number of magazines that have a woman on the cover. She goes to the book store and looks at the rack of magazines. There are at least 30 magazines. Amy looks at the covers of 20 magazines. Did Amy get a good sample?

Ⓐ Yes, because she looked at all of the magazines.
Ⓑ Yes, because she got a sample of more than half of the magazines.
Ⓒ No, because she did not looked at all of the covers.
Ⓓ No, because she looked at less than half of the magazines.

10. Travis is putting together outfits for work. He has 4 red shirts, 8 blue shirts and 7 white shirts. He has 6 pairs of khaki pants. What percent of Travis' possible outfits have blue shirts? Round your answer to the nearest whole number.

Ⓐ 8%
Ⓑ 43%
Ⓒ 42%
Ⓓ 48%

11. Travis is putting together outfits for work. He has 4 red shirts, 8 blue shirts and 7 white shirts. He has 6 pairs of khaki pants. What percent of Travis' possible outfits do not have red shirts?

12. **Scientists were concerned about the survival of the Mississippi Blue Catfish, so they collected data from samples of this species of fish. The scientists captured the fish, measured them, and then returned them to the lake from which they were taken. Select the correct numbers for different ranges of lengths.**

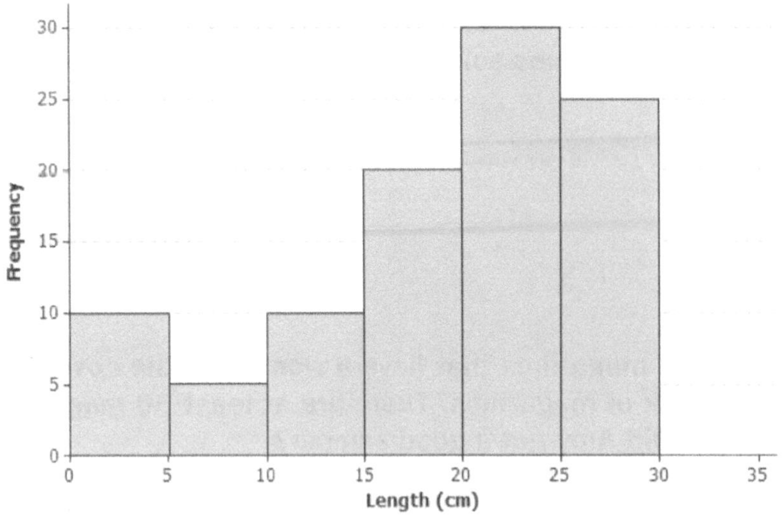

	5	10	20	25	30
0 - <5 cm	○	○	○	○	○
5 - <10 cm	○	○	○	○	○
10 - <15 cm	○	○	○	○	○
15 - <20 cm	○	○	○	○	○
20 - <25 cm	○	○	○	○	○

13. **Each statement below describes a sample of data. Select all the statements that describe the situations that would result in non-biased or equal representation of the data.**

Ⓐ Bella randomly called 100 phone numbers in her town to survey them about their shopping habits. 20 people chose to participate.
Ⓑ Heather climbed 11 trees throughout each forest in her county. Each forest is the same size.
Ⓒ Arujun bought 2 eggs from each farmer at the farmer's market. Each farmer had an equal number of eggs.
Ⓓ Allen put the names of all the cities in a state into a jar. Then he drew out 32 names from the jar.
Ⓔ Kiera polled 12 people at the town senior center.

Do NOT write your answers in this book. To open the answer sheet, scan the QR code or visit ***lumoslearning.com/a/6m040***

Chapter 5 → Lesson 6: Describing the Nature

1. **A new sandwich shop just opened. It is offering a choice of ham sandwiches, chicken sandwiches, or hamburgers as the main course and French fries, potato salad, baked beans or coleslaw to go with them. How many different ways can a customer have a sandwich and one side order?**

   Ⓐ 6
   Ⓑ 8
   Ⓒ 10
   Ⓓ 12

2. **Karen has a set of numbers that she is working with {6, 14, 28, 44, 2, −6}. What will happen to the mean if she adds the number −8 to the set?**

   Ⓐ The mean will decrease.
   Ⓑ The mean will increase.
   Ⓒ The mean will stay the same.
   Ⓓ It cannot be determined.

3. **Susan goes to the store to buy supplies to make a cake. What is the average amount that Susan spent on each of the ingredients she bought?**
   {$4.89, $2.13, $1.10, $3.75, $0.98, $2.46}

   Ⓐ $5.22
   Ⓑ $2.55
   Ⓒ $2.25
   Ⓓ $2.50

4. **Robert had an average of 87.0 on his nine math tests. His scores on the first eight tests were {92, 96, 83, 81, 94, 78, 93, 70}. What score did Robert receive on his last test?**

   Ⓐ 90
   Ⓑ 89
   Ⓒ 96
   Ⓓ 87

5. **Terry is playing a game with his brother and they played 5 times. The median score that Terry got was 37. The range of Terry's scores was 40. What was the lowest and highest score that Terry got?**

Ⓐ 37 and 77
Ⓑ 10 and 50
Ⓒ 17 and 57
Ⓓ Not enough information is given.

6. **Marcus wants to find out how many people go to the zoo on Saturdays in August. The zoo is open for 7 hours and there are 4 Saturdays in August. He counts the number of people who enter the zoo for two hours one Saturday. Will Marcus get an accurate idea of how many people go to the zoo?**

Ⓐ Yes because he will get a good representative sample.
Ⓑ No because he will not get enough of a representative sample.
Ⓒ Yes because he can assume that two hours is enough time to count the number of people.
Ⓓ No because he would need to be there for less time.

7. **Carl surveys his class to find out how tall his classmates are.**
How many classmates did Carl survey?

Height	Number of Classmates
4'4" - 4'8"	1
4'9" - 5'	4
5'1" - 5'5"	7
5'6" - 5'11"	14
6' - 6'4"	2

Ⓐ 27
Ⓑ 28
Ⓒ 14
Ⓓ 24

8. Julie is making a quilt. She uses 7 different patterns for her quilt squares.

What percentage of Julie's quilt is not flowers or butterflies?

Squares	Number
Solid	15
Stripe	22
Flower	8
Plaid	12
Zig-Zag	13
Circles	10
Butterflies	6

Ⓐ 80%
Ⓑ 72%
Ⓒ 84%
Ⓓ 83%

9. Carl surveys his class to find out how tall his classmates are.

How many classmates are taller than 5 feet?

Height	Number of Classmates
4'4" - 4'8"	1
4'9" – 5'	4
5'1" – 5'5"	7
5'6" – 5'11"	14
6' – 6'4"	2

Ⓐ 23
Ⓑ 14
Ⓒ 27
Ⓓ 16

10. Travis is putting together outfits for work. He has 4 red shirts, 8 blue shirts and 7 white shirts. He has 6 pairs of khaki pants. What percent of Travis' possible outfits do not have red shirts?

Ⓐ 79%
Ⓑ 90%
Ⓒ 78%
Ⓓ 15%

11. What is the mean of these numbers? 5, 4, 3, 7, 8, 9. Circle the correct answer choice.

Ⓐ 6
Ⓑ 5
Ⓒ 7
Ⓓ 3

12. Dana's recent math exams grades were: 67, 62, 70, 69, 90, 93, 95, 88, and 93. What is Dana's median score? Circle the answer that represents the median score.

Ⓐ 88
Ⓑ 93
Ⓒ 67

Do NOT write your answers in this book. To open the answer sheet, scan the QR code or visit ***lumoslearning.com/a/6m041***

Chapter 5 → Lesson 7: Context of Data Gathered

1. Bella recorded the number of newspapers left over at the end of each day for a week. What is the average number of newspapers left each day?

3, 6, 1, 0, 2, 3, 6

Ⓐ 2 newspapers
Ⓑ 3 newspapers
Ⓒ 4.3 newspapers
Ⓓ 6 newspapers

2. Recorded below are the ages of eight friends. Which statement is true about this data set?

13, 18, 14, 12, 15, 19, 11, 15

Ⓐ Mean > Median > Mode
Ⓑ Mode < Mean < Median
Ⓒ Median > Mode < Median
Ⓓ Mode > Mean > Median

3. The line plot below represents the high temperature in °F each day of a rafting trip. What was the average high temperature during the trip?

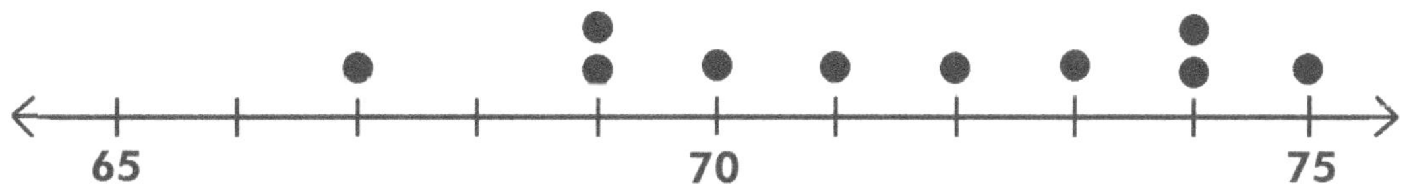

Ⓐ 70°F
Ⓑ 71°F
Ⓒ 71.4°F
Ⓓ 72°F

4. **What is the mean absolute deviation of the grades Charlene received on her first ten quizzes: 83%, 92%, 76%, 87%, 89%, 96%, 88%, 91%, 79%, 99%.**

Ⓐ 4.5%
Ⓑ 5.4%
Ⓒ 88%
Ⓓ 88.5%

5. **The following numbers are the minutes it took Nyak to complete one lap around a certain dirt bike course. What is the mean absolute deviation of this data set?**

12.3, 12.6, 12.2, 10.9, 11.3, 10.3, 11.7, 10.7

Ⓐ 0.7 minutes
Ⓑ 0.75 minutes
Ⓒ 11.1 minutes
Ⓓ 1.5 minutes

6. **The local Akita Rescue Organization has nine Akitas for adoption. Their weights are 82 lb, 95 lb, 130 lb, 112 lb, 122 lb, 72 lb, 86 lb, 145 lb, 93 lb. What are the median (M), 1st Quartile (Q1), 3rd Quartile (Q3) and Interquartile Range (IQR) of this data set?**

Ⓐ M = 95, Q1 = 84, Q3 = 126, IQR = 42
Ⓑ M = 95, Q1 = 85, Q3 = 126, IQR = 145
Ⓒ M = 95, Q1 = 84, Q3 = 145, IQR = 42
Ⓓ M = 95, Q1 = 126, Q3 = 145, IQR = 42

7. **Monica is keeping track of how much money she spends on lunch each school day. This week Monica spent $5.20, $6.50, $3.75, $0.75, and $4.15. What is the median (M) and Interquartile Range (IQR) of Monica's lunch expense?**

Ⓐ M = $4.15, IQR = $1.45
Ⓑ M = $4.07, IQR = $1.45
Ⓒ M = $4.07, IQR = $2.25
Ⓓ M = $4.15, IQR = $3.60

8. **Monica's friend Anna is also tracking her school lunch expenses. This week Anna spent $2.60, $0, $7.00, $4.40, $3.75. What is the difference between Monica's and Anna's average daily lunch expense?**

Monica: $5.20, $6.50, $3.75, $0.75, and $4.15

Ⓐ $0.40
Ⓑ $0.52
Ⓒ $0.60
Ⓓ $1.50

9. Zahra and Tyland are very competitive. Each have recorded their basketball scores since the beginning of the season. Which data set has a greater Interquartile Range (IQR)?

Name	Scores
Zahra	8, 10, 3, 12, 14, 11
Tyland	6, 7, 18, 12, 4, 9

Ⓐ Zahra has the higher IQR of 4.
Ⓑ Zahra has the higher IQR of 10.5.
Ⓒ Tyland's IQR is 2 points higher than Zahra's.
Ⓓ Tyland and Zahra have the same IQR of 5.

10. Jamila records the number of customers she gets each day for a week: 20, 42, 15, 23, 53, 62, 58. What is the mean absolute deviation of this data?

Ⓐ 16.9 customers
Ⓑ 19 customers
Ⓒ 39 customers
Ⓓ 47.5 customers

11. Frank played 9 basketball games this season. His scores were 15, 20, 14, 36, 20, 10, 35, 23, and 24. What is the median of Frank's scores? Write the answer in the box.

12. Gordan has the following data: 13, 19, 16, z, 17
If the mean is 16, which number could be z? Circle the correct answer.

Ⓐ 11
Ⓑ 15
Ⓒ 18

Do NOT write your answers in this book. To open the answer sheet, scan the QR code or visit ***lumoslearning.com/a/6m042***

Chapter 5 → Lesson 8: Relating Data Distributions

1. The histogram below shows the grades in Mr. Didonato's history class.

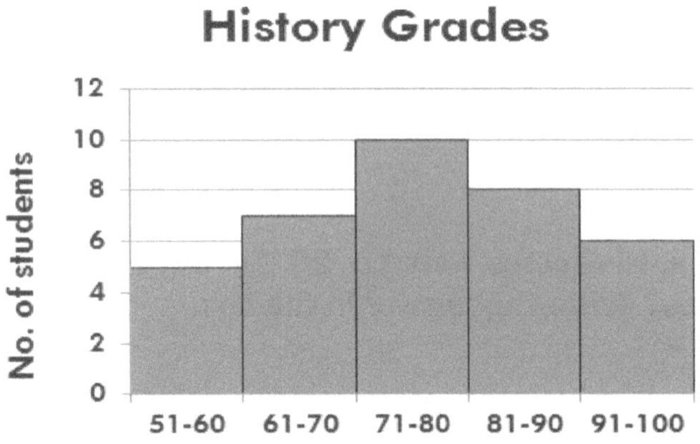

Which of the following statements is true based on this data?

Ⓐ The mode score is 75%.
Ⓑ More than half the students received 81% or higher.
Ⓒ Ten students received a 70% or lower.
Ⓓ There is not enough information to determine the median score.

2. **The bar graph below shows the different types of shoes sold by *Active Feet.***

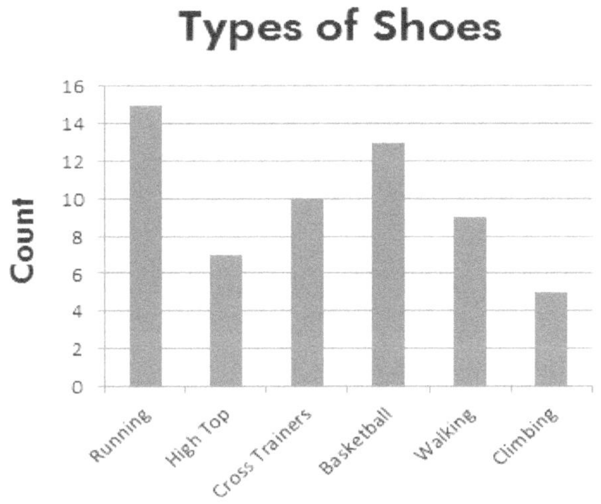

Which of the following statements is true based on this data?

Ⓐ The store carries more types of basketball and climbing shoes combined than types of running shoes.
Ⓑ The store sells more running shoes than any other shoe type.
Ⓒ The store has more types of walking shoes than cross trainers.
Ⓓ The store earns more money selling basketball shoes than walking shoes.

3. **Milo and Jacque went fishing and recorded the weight of their fish in the line plots below.**

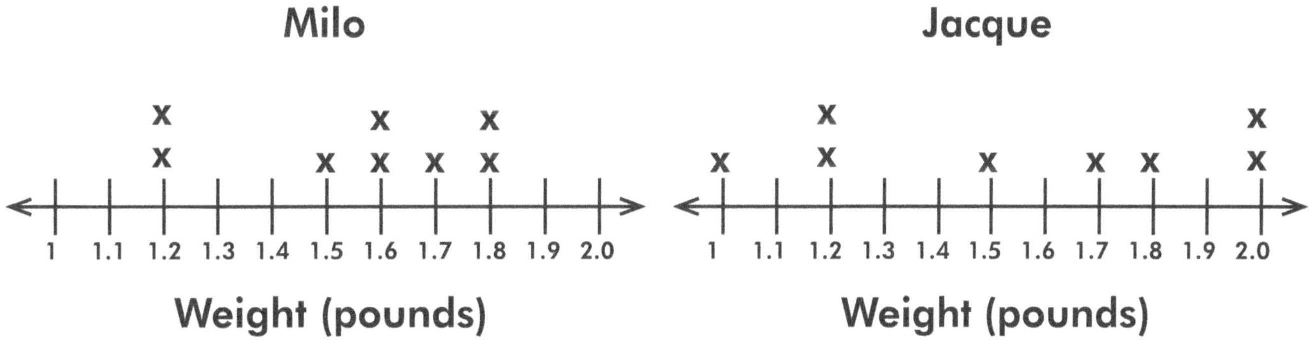

Which of the following statements is true based on this data?

Ⓐ The weights of Milo's fish have more variability.
Ⓑ The average weight of Jacque's fish is greater than the average weight of Milo's fish.
Ⓒ The weights of Jacque's fish have a greater mean absolute deviation.
Ⓓ Milo caught more fish by weight than Jacque.

4. **The frequency table below records the age of the student who attended the dance.**

Age	Frequency
11	17
12	20
13	9
14	8
15	12
16	4

Which of the following statements is true based on this data?

Ⓐ 80 students attended the dance.
Ⓑ The mean age is 12.86
Ⓒ If 3 more 16-year-olds arrive, the median age would increase.
Ⓓ More than half the students are older than the mean.

5. **For twelve weeks Jeb has recorded his car's fuel mileage in miles per gallon (mpg). He has plotted these fuel mileages below in a box plot.**

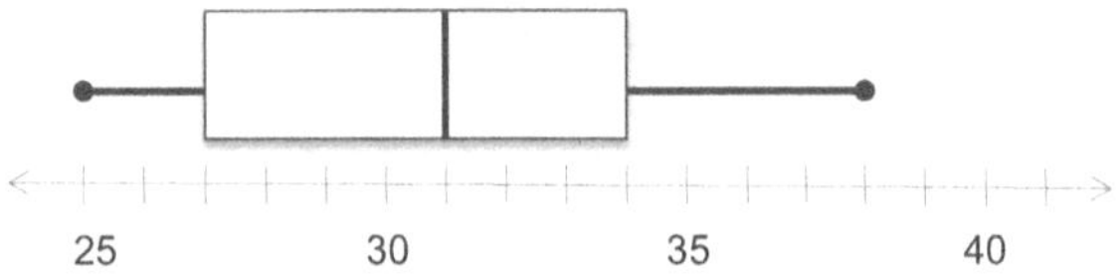

Which of the following statements is true based on this data?

Ⓐ The best fuel mileage recorded was 34 mpg.
Ⓑ Half of the data collected was between 27 mpg and 34 mpg.
Ⓒ The median fuel mileage was 30.5 mpg.
Ⓓ Three data points lie in the Interquartile range.

6. ***"Forever Green"* recorded the heights of the pine trees sold this month in the box plot shown below. There are 15 data points in Quartile 3.**

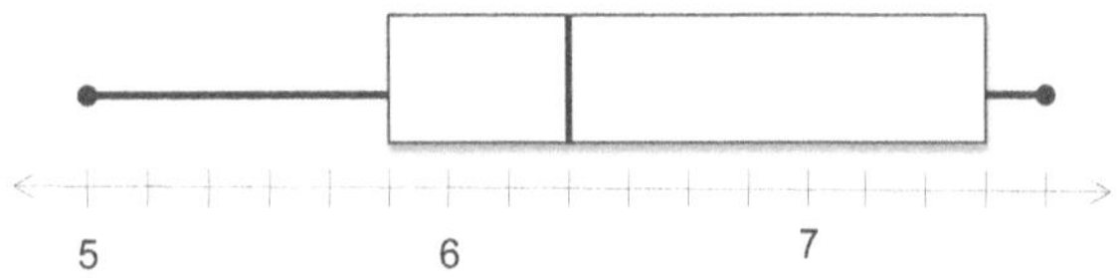

Which of the following statements is true based on this data?

Ⓐ About 25% of the trees were 7 feet 3 inches or taller.
Ⓑ There are more data points in Quartile 2 than Quartile 3.
Ⓒ The median height is 6 feet 2 inches.
Ⓓ "Forever Green" sold a total of about 60 trees this month.

7. **As part of a math project, each student in Ms. Lauzon's class recorded the number of jumping jacks they could perform in 30 seconds.**

Stem	Leaf
3	0 0 2 4 7 7
4	1 1 4 7 9
5	2 3 4 4 6 6 7
6	0 0 1 1 3 8 8
7	1 2
3|0 means 30	

Which of the following statements is true based on this data?

Ⓐ There are 21 students in Ms. Lauzon's class.
Ⓑ The median number of jumping jacks performed was 53.
Ⓒ The range of jumping jacks is 42.
Ⓓ Seven people performed less than 40 jumping jacks.

8. **Tamryn and Clio both make beaded necklaces. The line plots below display the number of beads on ten necklaces each from Tamryn and Clio.**

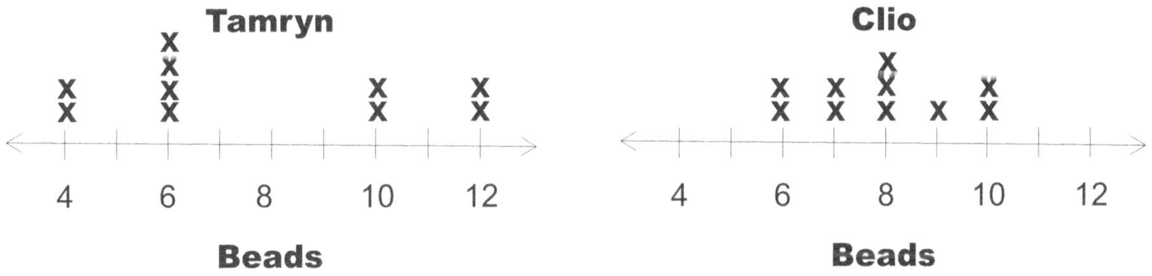

Which of the following statements is true based on this data?

Ⓐ If Tamryn added an eleventh data point of 10, the mean would not change.
Ⓑ On an average, Tamryn uses more beads than Clio.
Ⓒ Clio has more variability in the number of beads she uses.
Ⓓ If Clio added an eleventh data point of 8, the median would stay the same.

9. **The histogram below shows the high temperature each day for a month.**

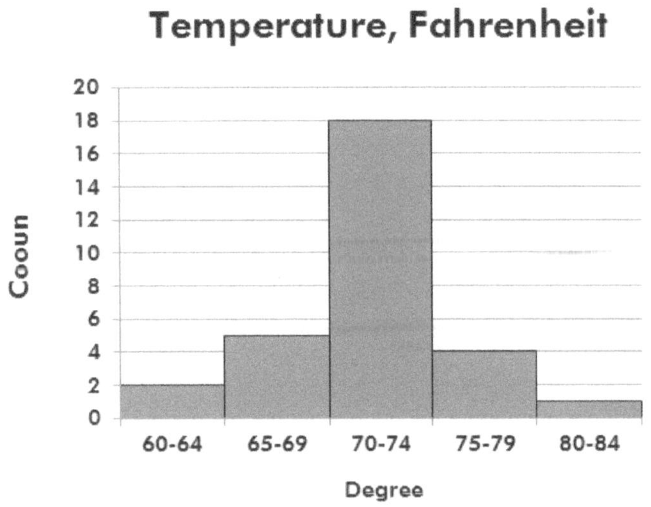

Which of the following statements is true based on this data?

Ⓐ More than half the days had a temperature of 70–74°.
Ⓑ 20% of the days had temperatures above 74°.
Ⓒ The average temperature was 72°.
Ⓓ There were fewer days below 70° than days above 74°.

10. **Marcel enjoys catching fireflies. The steam-and-leaf plot shows the number of fireflies Marcel caught on several nights.**

Stem	Leaf
0	4 6 9
1	0 3 4 7 7 7 8
2	0 0 2 2 5
3	1 4
1|3 means 13	

Which of the following statements is true based on this data?

Ⓐ The median is greater than the mean.
Ⓑ The median number of fireflies caught was 17.
Ⓒ The range of fireflies is 34.
Ⓓ Marcel caught fireflies on 14 nights.

11. **The students in one math class were asked how many brothers and sisters (siblings) they each have. The plot below shows the distribution of the data (answers from the students). Circle the blue box that represents the number of siblings had by the most students in this class.**

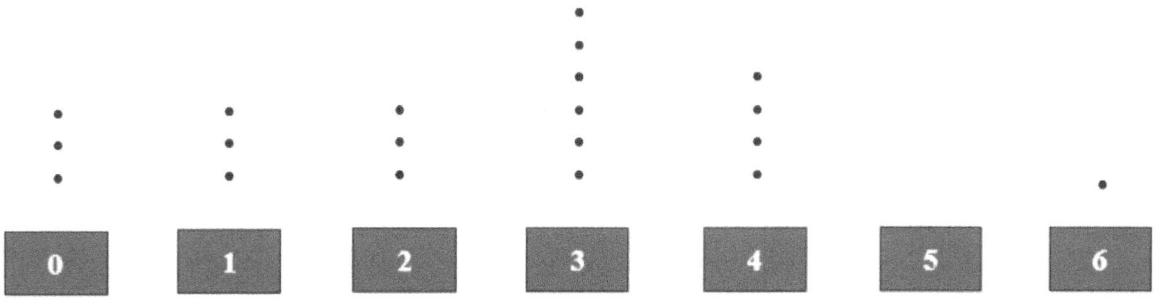

12. **Three volleyball teams each recorded their scores for their first 5 games. Use each team's scores below to determine the missing value.**

	10	14	17
Team 1: 15, 16, t, 17, 12, Mode = 17, What is t?	○	○	○
Team 2: 7, 14, r, 16, 13 Mean = 12, What is r?	○	○	○
Team 3: 11, 7, 19, 14, z, Median = 14, What is z?	○	○	○

End of Statistics & Probability

Michigan Student Test of Educational Progress Test Prep: 6th Grade Math Practice Workbook and Full-length Online Assessments: M-STEP Study Guide

Contributing Editor - Renee Bade
Contributing Editor - Kimberly G.
Executive Producer - Mukunda Krishnaswamy
Program Director - Anirudh Agarwal
Designer and Illustrator - Sowmya R.

ISBN 13: 978-1959697794

Printed in the United States of America

CONTACT INFORMATION

LUMOS INFORMATION SERVICES, LLC

PO Box 1575, Piscataway, NJ 08855-1575
www.LumosLearning.com
Email: support@lumoslearning.com
Tel: (732) 384-0146
Fax: (866) 283-6471

Lumos Learning
Step Up Your Skills

What if I buy more than one Lumos tedBook?

Step 1 ⟶ Visit the link given below and login to your parent/teacher account

www.lumoslearning.com

Step 2 ⟶ Go to the **"My tedBooks"** section and place the book access code and submit (See the first page for access code).

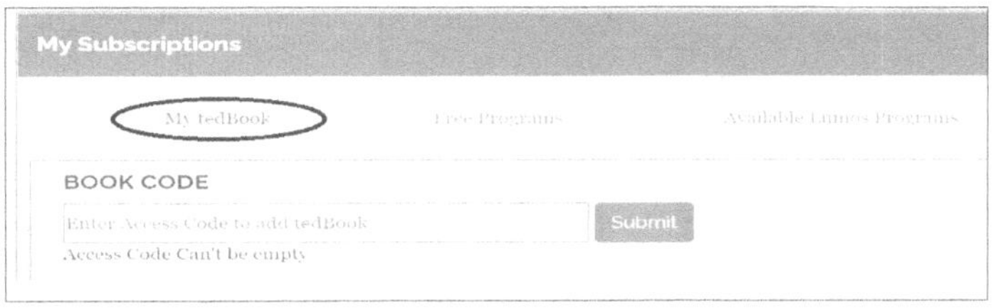

Step 3 ⟶ Add the new book

To add the new book for a registered student, choose the '**Student**' button and click on submit.

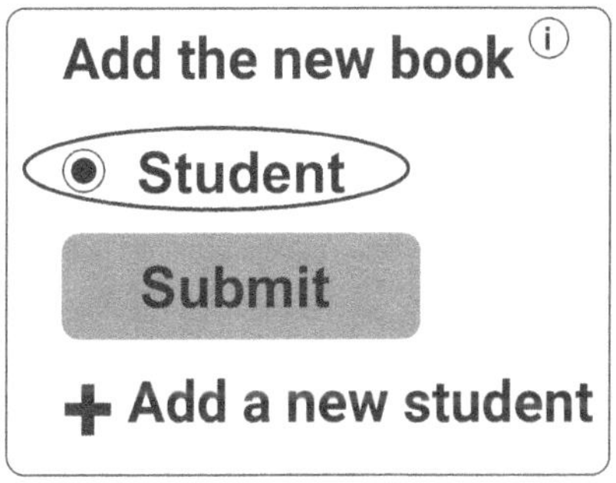

To add the new book for a new student, choose the '**Add New Student**' button and complete the student registration.

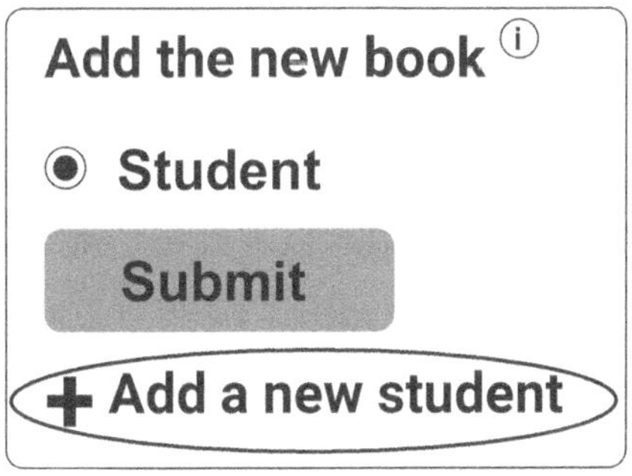

Also Available

Made in United States
Cleveland, OH
12 April 2026

35716958R00092